Insights into Chinese Agriculture

Yunhua Zhang

Insights into Chinese Agriculture

上海遠東出版社

Springer

Yunhua Zhang
Beijing, China

Translated by Curtis Evans

ISBN 978-981-13-1049-2 ISBN 978-981-13-1050-8 (eBook)
https://doi.org/10.1007/978-981-13-1050-8

Jointly published with Shanghai Far East Publisher, Shanghai, China

The print edition is not for sale in China Mainland. Customers from China Mainland please order the print book from: Shanghai Far East Publisher.

Library of Congress Control Number: 2018945096

Translation from the Chinese language edition: *Insights into Chinese Agriculture* by Curtis Evans,

Printed on acid-free paper

This Springer imprint is published by the registered company Springer Nature Singapore Pte Ltd.
The registered company address is: 152 Beach Road, #21-01/04 Gateway East, Singapore 189721, Singapore

Foreword I

One China: Building Agricultural Power—The Burden is Heavy and the Road is Long

Comrade Zhang Yunhua has long been engaged in researching agricultural and rural economic policy at the Development Research Center of the State Council. He has his own independent analysis and understanding of these issues. *Insights into Chinese Agriculture* is the crystallization of his many years of thought.

Farming is the source of food and the basis for the existence and development of humanity. In the process of human civilization, the development of society has demanded that more and more people divorce themselves from agricultural production, even to the point of breaking away from material production altogether. Nevertheless, the fundamental uniqueness of agriculture is that it is irreplaceable. No matter how rapidly technology progresses, industrialization cannot substitute for it. The production of foodstuffs and a major portion of the raw materials used in light industries as of yet can only be provided by agricultural production. No matter how the economy develops, if people want to survive then they will have to eat and wear clothing; there is really no way that this will change. "The masses regard food as their prime want."[1] If farming does not flourish, then people's existence cannot be assured and society will have no basis for further development.

Over the last thirty-five years of China's policy of reform and opening to the outside world, China has entered the middle stage of industrialization and rapid urbanization. Agriculture itself has also entered a new stage of development. It makes up a lower and lower proportion of China's gross domestic product (GDP). The increases in farmers' incomes rely less on farming and more and more on non-agricultural industries.

---

[1]This is an idiom slightly paraphrased from the *Hanshu* (*History of the* [*Former*] *Han Dynasty*) by Ban Gu (A.D. 32–92) and Ban Zhao (c. A.D. 48–c.112), one of the twenty-four official dynastic histories.—trans.

In this new stage of development, agriculture has not been weakened, but to the contrary has been given a new embodiment. Along with the increase in China's population and with the rise in living standards, society has increased its demands for higher quality and a greater variety of farm commodities. If farming is not able to satisfy these demands, economic development will be adversely affected and result in major problems. Although China can utilize international agricultural markets to adjust its domestic surpluses and deficiencies, as an economic power, China must mainly rely on its own domestic farm commodities to satisfy supply and demand. Farming not only merely provides our needs for commodities and extensive employment, but it also provides an excellent ecological system and serves diverse functions relating to life, education, and culture. It is clear that the higher the level of economic development, the more advanced society is and the more diverse agriculture becomes. In appraising it, one not only merely looks at economy and technology, but also at safety, the environment, and human culture. Therefore, it is necessary to transform its concepts. Agriculture must assume greater and more persistent responsibility than it has at any time in the past. It is becoming increasingly important and irreplaceable, given that it has broad room to grow.

Since 2004, the central authorities have successively issued twelve *No.1 Central Documents*[2] regarding the "Three Rural Issues,"[3] formulating a series of policies to strengthen agriculture and benefit farmers. These documents have resulted in great changes for the better for farming and rural areas: The production of food has increased steadily, agricultural restructuring has proceeded at a steady pace, and farmers' incomes have continued to rise. Nonetheless, the overall basis of Chinese agriculture is still very weak; in the process of developing it, some new problems have emerged that are worth noticing. Now, the prices of some domestic and international farm commodities have been turned upside down and domestic prices of some important commodities have started to exceed the CIF dutiable value [i.e., the cost, insurance, freight, and value on which custom duties are paid] of similar foreign products.

The "ceiling effect" on the price of domestic farm products has started to manifest itself. That is because as farm labor has become more and more expensive, rent has risen higher and higher; investments in agricultural manufacturing are increasing, such as farm machinery, chemical fertilizers, agrochemicals, and agricultural plastic sheeting used for growing plants under cover. As a result, production costs have quickly escalated. It is ever more clear that the rise in the cost of farm products is causing an economic squeeze to be felt.

When China entered the World Trade Organization (WTO), we promised to do something about subsidies, for example, Amber Box subsidies, which are those that

[2]*No.1 Central Document* is the first document of the year issued by the Central Committee of the Communist Party of China (CPC); these often allude to the importance placed on agriculture.—trans.

[3]The "Three Rural Issues" are those of agriculture, rural areas, and farmers.—trans.

distort production or trade.[4] We promised to only institute subsidies that do not exceed 8.5% of the value of agricultural output. Now, some of our farm production and price subsidies have started to draw near to that "Amber Line." Under present trade protection policies, if our nation's farm commodities hit this 8.5% ceiling and price-driven imports continue to increase, it will be difficult to avoid shocks to Chinese agriculture caused by imports. There has been a long-term problem of over-exploitation and over-use of agricultural resources because we have severely overdrawn our ecological account in order to increase the volume of farm products. In some places, there is land that is impossible to farm and water that cannot be used for irrigation. The old environmental resource debt has not been repaid, and new such debt has continued to be run up. Red warning lights have started flashing that tell us of danger to our natural resources and the environment. One could say that the development of agriculture is confronted with the double whammy of a price ceiling and a price floor effect; at the same time, it also faces the flashing of environmental warning lights and the approaching "Amber Line" of WTO's Amber Box.

Facing these great changes and challenges, there is only one way out: China must quickly transform its approaches to agricultural development. How can this be done? In the past, such development depended to some degree on exploiting resources and the environment as much as possible, mainly seeking growth in the volume of farm output. Nonetheless, now we must shift to lay equal stress on quantity and efficient quality. China is such a large country that we must not ignore the quantity of production. We much guarantee the stability of output of major agricultural products. With this as its foundation, we must attach more importance to promoting the quality and efficiency of agricultural development. We must lay more stress on increasing the power of our agriculture in international competition, technological innovation, and sustainable development.

In December 2013, General Secretary Xi Jinping emphasized at the central rural work conference: If China wants to be strong, agriculture must be strong. If China wants to be beautiful, the countryside must be beautiful. If China wants to get rich, the farmers must get rich. Strengthen farming. Beautify the countryside. Enrich farmers. These are lofty concepts and grand propositions with abundant connotations. This is a matter of realizing the dream of creating an affluent society; this is a matter of realizing the Chinese Dream about the "Three Rural Issues." All of these

[4] In WTO terminology, subsidies of agriculture and agricultural commodities are categorized as belonging to one of three different-colored "boxes." Green Box: contains unlimited government subsidies that do not distort trade (e.g., research funding, inspection programs, disaster relief, farmer training). Blue Box: contains tolerated aid, i.e., subsidies that do not increase as production increases (e.g., subsidies linked to acreage or number of animals). Blue Box subsidies tend to be highly controversial. Amber Box: contains aid to be avoided and reduced, and includes almost all domestic support measures considered to distort production and trade (i.e., everything not included in the Blue Box or Green Box). The Amber Box contains subsidies for price supports and subsidies directly linked to production (e.g., seeds, fertilizers, irrigation subsidies), export subsidies, cheap loans, etc.—trans.

concepts are providing the basic direction for successfully dealing with the "Three Rural Issues" in the new era.

Granaries must be full, food and clothes must be ample, and all under Heaven must be peaceful: These have been the enduring and unchanging laws of society since ancient times. The genuinely powerful countries of the world are those that do not have soft underbellies. They are all able to resolve the problem of feeding themselves.

First of all, in strengthening agriculture, it is necessary that we take the initiative to ensure food security at all times. The major objective of our country's agricultural development is to guarantee that all people can not only "eat their fill," but also "eat well."

In strengthening agriculture, it is necessary to guarantee, "from farm to dinner table," the quality and safety of farm products and foodstuffs. The fundamental realities of our country are that there are too many people, too little land, and water is in short supply.

In strengthening agriculture, it is necessary to do the following: Give prominent position at all times to the preservation and conservation of resources in farm development. Establish and strengthen an extremely strict protective system of farmlands and an extremely strict system of economical land use. Develop husbandry that recycles, is ecological, and is conservation-oriented. In addition, form a development style that both conserve resources and are environmentally friendly. To equip farms with modern material conditions and to transform them with modern science and technology are important hallmarks of the modernization of agriculture. They are weak links in our country's agriculture that urgently require strengthening.

In strengthening agriculture, it is necessary to put great effort into implementing a scientifically and technologically strong strategy; to deepen reforms of the systems of agronomic science and technology; and to accelerate the construction of new innovative marketing systems. In the process of rapid urbanization, and in the resultant wake of large numbers of rural youths and adults migrating to cities to become laborers or to engage in trade, the quality of the farm labor force has revealed a structural decline. That is to say, the outflow of talent from rural areas is severe, manifesting prominent problems of "the hollowing out of villages," "farming becoming merely a sideline occupation," and "the aging of the farming population."

In strengthening agriculture, in order to solve the problem of "Who will farm the land?", it is necessary to guide farmers to voluntarily transfer their rights to rural contracted land management in order to encourage the development of diverse forms of operations on an appropriate scale. There needs to be professional corporate farming, management by multi-household enterprises, family farms, specialized cooperatives of farmers, etc. We also need to exert great efforts to train new professional farmers who are well educated, technologically sophisticated, and good at management. There should be farmers who take the initiative in entering the marketplace and an agriculture that takes the initiative to make this transition to the market. But this should not be a halfhearted marketization reformation.

In strengthening agriculture, it is necessary to act decisively to: unswervingly do away with all unreasonable restrictions and discrimination that hobble farmers; relinquish controls and let farmers audaciously force their way into markets; and genuinely enable markets to allocate resources. Cultivating the land is an industry that relies to a high degree on natural resources. In a country with too many people and too little land, it is difficult to greatly increase farm labor productivity. It is hard for small-scale farmers to deal with rising production costs. Farming's economic benefits to society are great, but its economic benefits to the individual farmer are small.

Finally, in strengthening agriculture, it is necessary to perfect policies for protecting farming. At present, although the upper limits of the Amber Box subsidies [i.e., those which have the effect of distorting production and trade] are already being approached, there are no limitations on Blue Box subsidies. We must increase policies that support farming and intensify efforts in the areas of infrastructure construction, insurance, poverty alleviation, and direct income support for farmers.

In this book, Comrade Zhang Yunhua analyzes and explains in simple terms the complicated problems of agriculture. He summarizes the achievements and existing problems of its development. He also reviews in summary fashion the unfolding process of the development of government policy regarding agriculture. Furthermore, he introduces fundamental and general principles of agricultural development. Finally, he explores the direction of future government policy in this area. The content of the whole book is professional and prescient, and is a work that allows one to understand Chinese agriculture in a relatively systematic manner. One hopes that the author will continue to diligently study and explore this subject in his future research.

Beijing, China
January 2015

Han Jun 韩俊
Director of the Rural Work Leading Group
Office of the Central Committee of CPC

Foreword II

It is not Easy to Understand Chinese Agriculture

What is agriculture like in China? In the minds of most urban Chinese, there appear images of ox-drawn plows and donkeys carrying sacks of grain on their backs. What they do not know is that in China the level of mechanization of plowing, cultivating, and harvesting has reached 59.48% (as of the end of 2013). That is, approximately 60% of farming practice has said good-bye to human and animal labor. The three staple grains (viz., wheat, rice, and corn) have reached levels of mechanization of 93.7, 73.1, 79.8%, respectively. It is thus clear that the first steps toward modern agriculture have already been taken. People have been given this false impression of Chinese agriculture by the literature, movies, and television shows; of course, it is also related to the fact that people in agriculture have not done enough to broadcast and disseminate the real picture. Therefore, in a useful attempt to do this, it is gratifying that Zhang Yunhua has written this popular work, *Insights into Chinese Agriculture*, in addition to conducting arduous scientific research as a researcher at the Rural Economy Research Department in the Development Research Center of the State Council.

Shortly before this manuscript went to press, I was invited to write a preface. Zhang Yunhua was the first doctoral candidate that I supervised, so it is my honor to write an introduction to his new book. To take advantage of this opportunity, I would like to talk about three aspects.

Agriculture and Civilization

According to the definition in the *Xiandai Hanyu Cidian* (*Modern Chinese Dictionary*), culture is "the sum of material and spiritual wealth created by mankind in the process of developing society and history." Within the broad framework of the concept, culture includes both production methods and lifestyles. China has just entered the middle stages of industrialization. Its modes of production have already been transformed. But even the white-collar class residing in the big cities has still not been able to fundamentally transform its lifestyle. Nothing can be done about this, just as a person cannot change the yellowness of his skin no matter how much coffee he drinks. Culture is deeply ingrained, right down to the marrow. It is a bit like blood types: A, B, O, AB, etc.; after one is born, one cannot change one's blood type, try as one might.

China has undergone an extremely long period as an agricultural society. Shennong[5] sampled hundreds of kinds of plants to discover which ones could be used for medicine. He also taught the clansman of his tribe to grow crops; this was the very beginning of Chinese agriculture. When this venerable gentleman sampled the myriad of plants, every time he tasted a bitter one that poisoned him, and he then drank some water that had tea leaves steeped in it to counteract the toxin. This fortuitous discovery of tea bequeathed us with two very deep-rooted customs that have become part of our culture.

One of these customs is the drinking of tea. It is hard to compare the varieties of Chinese tea with those of any other country. We classify them by color: green tea, red tea, light oolong tea, black tea, white tea, yellow tea, floral tea, etc. There is every kind of tea that one could wish for. Our red tea is called "black tea" in English; this seems preposterous to us, since we actually have *heicha*—a true, black "black tea." Some people have argued that the etymology of the word "China" traces back to the Chinese word *cha*, meaning "tea" and thus came to represent the country that first produced it.[6] This is understandable because the trade in tea had a tremendous impact on Western countries. Some people even think that the real cause of the Opium Wars was that because China made a great deal of money in the tea trade, Western countries faced an imbalance in silver reserves. Britain then chanced on opium to sell to Chinese as a way to redress this trade imbalance.

The other custom is that we Chinese very seldom eat bitter foods (the eating of bitter melon is an anomaly; in fact, it was imported from India only during the Late Ming period [i.e., the early seventeen century]. Precisely because eating bitter herbs poisoned the venerable Shennong, it is in Chinese genes to refuse bitter tastes and thus we reject bitter-tasting food. Farming culture is ubiquitous.

Thousands upon thousands of years of farming history have left the Chinese people, who are descendants of the Fiery Emperor [Yandi] and the Yellow Emperor [Huangdi],[7] with an enormous cultural heritage, both tangible and intangible, material and spiritual. In an agricultural society, the most important thing is the

---

[5]I.e., the Devine Husbandman, a cultural hero from the dawn of Chinese history; he is mentioned further in Section III of the prologue of this book.—trans.

[6]The more commonly accepted etymology for "China" is that the name derives from the Persian *chīnī* "relating to China," in reference to where Chinese porcelain originated. Some scholars think that the Persian "*chīn*" ultimately derives from the *qin* in "Qin dynasty," the distant land of silks, porcelain, and tea being referred to as the Land of the Qin.—trans.

[7]These are the two earliest and greatest culture heroes of China, their origins lost in the mist of time, preceding even the semi-mythical Xia dynasty. "Emperor" is an honorific bestowed by later generations; Yandi and Huangdi would have been chieftains of their respective clans.—trans.

farming season. Since the Qin and Han dynasties, the twenty-four solar divisions that divide the year into twenty-four equal periods, based on the farming seasons of the Yellow River drainage area, have rationally been summed up in the following poem[8]:

> *Spring: There are the periods of Beginning of Spring, Rain Water, Waking of Insects, Spring Equinox, Pure Brightness, and Grain Rain.*
>
> *Summer: They are then linked with those called Beginning of Summer, Lesser Fullness [of Grain], Grain in the Ear, Summer Solstice, Lesser Heat, and Greater Heat.*
>
> *Autumn: These are then followed by the periods of Beginning of Autumn, End of Heat, White Dew, Autumnal Equinox, Cold Dew, and Frost's Descent.*
>
> *Winter: Next there follows Beginning of Winter, Lesser Snow, Greater Snow, Winter Solstice, Lesser Cold, and finally Great Cold.*
>
> *In the first half of the year, the start of the divisions usually falls on the sixth day of each lunar month and on the twenty-first day of each such month.*
>
> *In the last half of the year, the start of the divisions usually falls on the eighth of each lunar month and on the twenty-third of each such month.*
>
> *Every lunar month is divided into two divisions [of about two weeks each],*
>
> *At most, this principle [of when the different divisions fall during the first and last halves of each month] has a yearly variance of only one or two days.*

As the old farm saying goes, "In sowing winter wheat, the period of the White Dew is too early while that of the Cold Dew is too late. The Autumn Equinox period is just the right time." When the twenty-four solar divisions were defined, each was associated with a specific farming task in mind. This is truly an extensive and profound system of farm culture. What is amazing is that there is no contradiction between modern science and this discovery by our ancestors; modern science has proven the scientific nature and precision of the ancient system.

More than 2,000 years ago, Confucius earnestly advised his disciples:

> To rule a country of a thousand chariots, one must dispatch business with dignity and good faith; be thrifty and love all men; and mobilize the labor of the common people only during the proper seasons [so as not to interfere with farming operations]. (The Confucian Analects, Xue'er 1.v)

From Confucius' point of view, to run a country of 1,000 war chariots required nothing more than these three things. It is important to notice that the last of these strictures is that the common people must be allowed by good government to do their farm work in the right season, neither too early nor too late. That is to say, corvée labor[9] must not be demanded during the farming season. Thus, by respecting

[8]This anonymous folk poem, known as *Ershisi jieqi ge* (*Song of the twenty-four solar terms*), was written as a heuristic aid to help farmers to understand the twenty-four solar periods that constitute the year according to the traditional Chinese calendar and to thus know when to plant, harvest, etc. It also cleverly tells what day of the month to expect the start of new periods.—trans.

[9]I.e., non-agricultural work that the government required of peasants as part of their taxes, usually twice annually.—trans.

the sanctity of the farming season, one is able to achieve harmony between man and nature, a concept that lies at the very center of Chinese agriculture. In fact, I view this core idea as both the point of departure and the culmination of all Chinese culture.

This core of Chinese culture may also be found in the Confucian Doctrine of the Mean; it is a doctrine of a farming society, so it naturally was born from farming culture. In fact, the principle of not violating the farming seasons was transplanted to the social realm, in order to offer guidance in how to conduct oneself in society and govern society.

> The Master [i.e., Confucius] said, 'The Doctrine of the Mean is the supreme virtue!' (The Confucian Analects, Yongye 6.xxvii)[10]

That is to say, the Doctrine of the Mean is the highest form of benevolence. In today's world, whether governing the country or conducting oneself in society, the Doctrine of the Mean is the highest principle that must be followed. From this, one can tell that farming and farm society have had an enormous influence on us.

Farming Culture, Industrial Culture, and Ecological Culture

The culture that is produced by a farming society is called farming culture and that produced by an industrial society is called industrial culture. Farming culture has already had more than 10,000 years of history, while industrial culture started in the mid-eighteenth century with the Industrial Revolution in England. While it only started three hundred years ago, which compared to the length of farming culture is nothing, nonetheless it has produced great societal wealth. In the *Communist Manifesto*, written by Karl Marx (1818–1883) and Frederick Engels (1820–1895), it explains: "The bourgeoisie, during its rule of scarce one hundred years, has created more massive and more colossal productive forces than have all preceding generations together."[11] But that monster, the Industrial Revolution, by seeking high profits, gobbled up all sorts of natural resources with a speed beyond imagination. Natural resources have continuously dried up and resulted in environmental problems. From a historical perspective, the relationship between man and nature has long been one of exploiter and exploited.

During this period of excessive exploitation, nature has punished humanity. Witness, for example, the extinctions of the ancient Mesopotamian cultures of Western Asia and that of the ancient desert kingdom of Loulan on the Silk Road in Xinjiang in China's western region. Yet, exacting this sort of penalty was an

[10]The Doctrine of the Mean is an important Confucian concept. The *Zhongyong* (*Doctrine of the Mean*) forms a chapter of the *Liji* (*Classic of Rites* or *Book of Rites*) and is a pillar of the Confucian canon.—trans.

[11]Friedrich Engels, ed., *Collected Works of Marx and Engels*, 1888.—trans.

extremely slow process and only happened in some certain corners of the world; these extinctions of ancient cultures did not affect the whole landscape of humanity and are largely unknown to most of the world. However, this is not the case anymore for industrial culture. At the same time, all industrialized countries have used excessive quantities of natural resources and have polluted the environment. As a result, negative effects have manifested themselves to some extent.

Rachel Carson's *Silent Spring* was published in 1962, arousing mankind's awareness of the need to protect the environment. All sorts of environmental organizations sprouted up one after another and spurred the United Nations to convene the United Nations Conference on the Human Environment (UNCHE) in 1972, in Stockholm, Sweden, with participating nations signing "the Declaration of the United Nations Conference on the Human Environment" [also known as the "Stockholm Declaration"]. This marked a global initiative to protect the environment. In March 1983, the United Nations used the term "sustainable development" for the first time. In November 1983, the United Nations established the World Environment and Development Committee. In June 1992, the United Nations Conference on Environment and Development (UNCED)[12] convened in Rio de Janeiro, Brazil, and adopted and signed important documents, including the *Rio Declaration on Environment and Development* and *Agenda 21*. With these, the concept of sustainable development became widely accepted around the world.

Environmental protection in China came into being in the 1970s. In 1979, the Chinese government promulgated the *Environmental Protection Laws of the People's Republic of China*, (*trial implementation*). In 1994, the Chinese government promulgated *China's 21st Century Agenda*. It was the world's first national-level "Twenty-first-Century Plan of Action."

In 2007, in the report of the 17th National Congress of the Communist Party of China (CPC), for the first time the concept of ecological culture was proposed to, "... construct ecological culture and take the basic shape of industrial structure, growth methods, and consumption patterns to conserve natural resources and protect the ecological environment."

In 2012, the report of the 18th National Congress of the CPC gave a full explanation of "greatly promoting construction of ecological culture," pointing out that "We should stick to the basic state policy of conserving resources and protecting the environment by adhering to the policy of prioritizing conservation, protection, and natural recovery. This will promote green development, recycling development, and low-carbon development, and thereby form green belts, industrial structures, modes of production, and lifestyles to conserve resources and protect the environment. This will result in reversing at the source the trend toward environmental deterioration and thereby create a good working and living environment for people, thus making contributions to global ecological security."

We have made clear the evolution of ecological culture. Now, how are we to understand ecological culture? Exactly as was reported in the 18th National

[12]This is also known as the Rio Summit or Earth Summit.—trans.

Congress of the CPC: To build an ecological culture first and foremost requires an emphasis on conservation of natural resources, environmental protection and on forming a green, recycling, and low-carbon development mode.

What is the essential nature of all this? From what I understand, it is harmony—that is, the harmony of man, nature, and society. Is this not in accord with the ideology that man is an integral part of nature? Is this not in keeping with the Doctrine of the Mean? Is this not the aforementioned core of Chinese traditional culture? Circling around and around, we mistakenly thought that we had moved far away, but actually we have only returned to our origins—yet not to the original point. Instead, we have spiraled upward! At this point, we cannot say that ecological culture is the highest form of farming culture. Nonetheless, we can say that ecological culture has fully absorbed agriculture, especially its core idea—the unity of man and nature (i.e., harmony). Agriculture has added the wings of industrial culture to let it soar upward. It sounds somewhat paradoxical, does it not?

Now, let us return to the main subject of "Insights into Chinese Agriculture," i.e., the understanding of agriculture. During the past stage, that of farming culture, agriculture was the largest industry. In the present stage, that of industrial culture, it has become a smaller, tertiary industry. In developed countries, farm production generally does not exceed 5% of total domestic production; in China, at present it does not exceed 10%.

In the emerging stage of ecological culture, the percentage of agriculture has begun to gradually increase. Of course, this increase cannot be seen if one uses traditional methods of accounting; but the increase comes from being part of other service industries such as the tourist, leisure, and health industries. These are all closely related to agriculture and are components of it. In other words, in the ecological culture phase, its concept has been broadened and enlarged. It no longer simply produces food, but it also provides products for entertainment and relaxation: It is not only an important production industry, but also a service industry. According to data published on Web sites of the Ministry of Agriculture, by the end of 2013 the total number of all categories of national recreation businesses under agricultural management had already exceeded 1.8 million and had received some 900 million guests, with business revenues of 270 billion RMB, bringing benefits to 29 million farmers. The number of people and the operating revenue increased at an annual rate of 15%.

If it continues at this rate, what will it be like after ten more years? Furthermore, China's rate of urbanization has already exceeded 50%. Over half of citizens now live in cities, suffering the hardships of living in "concrete jungles"; what is to be done about this? Move agriculture to the cities! Of course, this does not mean producing crops in the cities, but it means planting trees and letting cities become "forest cities" in order to reduce some of the negative effects of the "concrete jungle." Forestry is one of the components of agriculture and if one includes it in the economic equation, then agriculture's proportion of the economy is even greater. By looking at it this way, contemporary agriculture is hardly what one has learned about in textbooks.

Knowledge About Agriculture Must Be Popularized

The above two questions offer a macroscopic perspective and are somewhat metaphysical. What will be talked about here is more concrete and closely related to our daily lives. That is, food, the thing we eat every day. We often hear this comment: In the past, people did not know that they would have anything to eat; now people do not know if they can eat anything that they have. This is a way of saying that there was not enough to eat before China's reform and opening to the outside world more than thirty years ago; but now that farm products have become more and more abundant, Chinese people are faced with the problem of food safety.

In particular, due to the rapid growth of the dairy industry, many problems have been exposed. In 2008, there were the melamine milk incidents[13]; this revealed to consumers the existence of problems of food safety not just in the dairy industry but also in the agriculture industry as a whole. A great many consumers to this day have not fully recovered their confidence in the safety of domestic farm products.

It should be mentioned that the issue of food safety can never be over-emphasized, because food is something that people directly ingest into their bodies; it bears on the health and even the life of every consumer. Since 2008, the government has enhanced the amendments to and the enforcement of food safety laws and regulations; it has strengthened supervision of the food supply chain from farm to dinner table. The food safety situation has already clearly been ameliorated. However, the problem lies in the fact that there is a considerable portion of consumers who do not pay attention to real food safety problems; rather, they pay attention only to the production process of farm commodities. For example, many people think that it is only the use of chemical fertilizers and agrochemicals that makes farm produce unsafe. There are manifold reasons for this kind of misunderstanding. For one thing, in recent years domestic and foreign scholars have opposed the extensive use of petroleum in farming and have firmly persisted in the traditional notion of production without the use of chemical fertilizers and agrochemicals. I call it the agricultural doctrine of "restoring the ancient ways." Their influence is an important reason that consumers misunderstand farm production. After all, people tend to believe the opinions of leading scholars. Another reason is that some of the media propagate the aforementioned scholars' perspectives and overstate the incidents related to the safety of farm products.

In that case, how are we to understand the use of chemical fertilizers and agrochemicals in farm production? There is a specific section in this book regarding this issue, so here I only want to discuss a famous theorem that is called the Theory of Induced Innovation, put forward by the agricultural economists Yujiro Hayami

---

[13]These scandals erupted in China when hundreds of babies were made ill and some died because of consuming melamine-contaminated milk formula. Drinking such milk interferes with kidney functions. Unscrupulous milk producers had added the chemical melamine to milk as a way to artificially boost protein levels; higher protein levels on tests conducted by wholesale milk buyers meant that sellers received higher prices for their milk.—trans.

(1932–2012) of Japan and Vernon Ruttan (1924–2008) of the USA. According to this theory, under conditions of a market economy and due to price mechanisms, farmers invariably choose resources that are plentiful to substitute for those that are scarce. Therefore, agricultural technology improvements are always advancing toward plentiful resources; farmers substitute the plentiful resources for those that are scarce. The use of chemical fertilizers and pesticides adheres to this principle.

In the 1960s, led by developed countries, the Green Revolution started in Southeast Asia. China, although relatively closed off from the outside world, urgently needed highly productive crops to replace traditional ones. That was because of the ever-increasing demand for food, following agrarian reforms during the process of industrialization. In 1958, the Green Revolution began. A starting point was development of "special southern dwarf hybrid" rice, a good strain of rice that was selectively bred by farmers themselves and was resistant to lodging [i.e., being blown down by strong winds]. The Green Revolution also stimulated innovation in agronomy. The approach of the Green Revolution was: The highly productive dwarf varieties of rice plants required high-efficiency chemical fertilizers and because high-yield varieties were continually being introduced, these in turn required fertilizers that were more and more efficient.

Since the 1970s, China had popularized the use of hybrid rice, but traditional (i.e., organic) farm fertilizers were not able to provide the requirements to grow this kind of new paddy rice. Therefore, the Chinese government went to great lengths to develop the chemical fertilizer industry. Hence, as Zhang Yunhua writes in Chap. 1, Section I, Subsection 4 of *Insights into Chinese Agriculture*: "… in 1973 and 1974 China started to employ its relatively scarce foreign currency reserves to purchase complete sets of chemical fertilizer production equipment from America, the Netherlands, France, and Japan. This was the famous '43 Program,' which referred to its total investment of $4.3 billion US dollars (at that time a huge sum of money) to import such equipment." In fact, we tightened our belt in order to produce chemical fertilizer. Without such fertilizer, hybrid rice could not be very productive; without supplying sufficient grain, national industrialization was a mute point. While chemical fertilizers promoted the growth of hybrid rice, they also promoted the growth of weeds; relying on manual work to pull up weeds was unsustainable. Ergo, herbicides were put into use. In the wake of the expansion of paddy rice, harmful insects became the greatest obstacle to the further improvement of yields; consequently, the application of pesticides was unavoidable.

Because of this, in the 1980s, China's agriculture thoroughly transformed from a traditional one to one that was petroleum-based. From the preceding analysis, one can see that this change was in accord with the Theory of Induced Innovation. It was inevitable in the transformation of China's economic and agricultural development; unless we wanted to go hungry, we had to modernize.

So, is petroleum-based farming harmful? Of course it is; farming development around the world has proved this. For example, extensive use of chemical fertilizer can result in encrustation of the surface soil and consequent decreased output; chemical fertilizer and pesticides lingering in the soil can pollute groundwater. farm products, etc. In practice, these problems all require step-by-step resolution. In reality, it would be impossible to stop the use of petroleum products and return to the days of traditional agriculture.

The data that Zhang Yunhua provides in this book shows that the use of fertilizer contributes around 50% to the food production volume in China. What I grasped from the data is that improved varieties, chemical use, and irrigation contribute to increased production in a ratio of 3:4:1. Without chemical fertilizer, pesticides, and herbicides, how could China sustain 1.3 billion people? Only if grain is supplied in sufficient quantities will we be able to economize on land use in order to have sufficient land to grow vegetables, fruit, and flowers, and thus provide an abundant life for everyone.

What we have to do is to eliminate the negative effects of petroleum-based farming on the technological level. For example, through precision agriculture, which controls the use of water and fertilizer, we can satisfy both the need for sufficient crops and at the same time reduce harmful residues as much as possible. In recent years, agriculture sectors have put into effect soil testing and formulae for applying fertilizer that are the first steps toward precision farming. This has proved very effective. In the long term, the adoption of biotechnology is the only way to solve the above-mentioned problems.

It is thus clear that it is not easy to give a comprehensive explanation of agriculture, and it is even more difficult to explain it to that half of the population that lives in urban areas. In any case, Zhang Yunhua has already made great strides in this direction.

Beijing, China
November 2014

Prof. Kong Xiangzhi 孔祥智, Ph.D.
Associate Dean, the School of Agricultural
Economics and Rural Development
Renmin University of China

Translator's Preface

It has been a pleasure to read and an honor to translate *Insights into Chinese Agriculture*, by Zhang Yunhua 张云华 (1976–). The story of China's agriculture, its history and development, and how it has managed to feed almost a billion and a half people are remarkable, complex, and important, and one that deserves to be better known in the Western world.

The author is a professional researcher of Chinese agricultural economy and writes from deep experience. He also draws on his own experience of growing up on a farm family in the dry, loess hills of Shanxi Province during the 1980s. His personal anecdotes help personalize the book, add humor, and assure the reader that his expertise stems from more than just theoretical learning. In fact, it is grounded in a deep, profound love and respect for the land, farming, and farmers.

The title of this book is apt for the Chinese reader, as the book is packed with insights into Chinese agriculture: information, statistics, theories, and opinions. But it is even more apt for the English reader, because in addition to the above there are many discerning observations about Chinese history, language, poetry, and culture.

The process of translation is similar to making sausage in that it is normally best kept from view of the consumer. But in order to help the reader more fully understand the book, the publisher asked that I offer the English reader a glimpse of the process. To do so, I have outlined below some aspects of that process that may be useful.

Overarching Aspects to Keep in Mind

First, this book was written for both general reader and specialist. As Mr. Zhang states in his epilogue, "For the last few years, it has been my dream to write a book on agriculture that is both popular and at the same time professional." Furthermore, it was written specifically for a Chinese audience, one that is aware of the cultural background, political and historical context, and linguistic conventions and tropes. Referencing is to China and Chinese; it was not written for an international audience. This one-size-fits-all, sino-centric approach presents challenges and opportunities to the English reader and to the translator.

The publishers wanted this translation to target those English readers, especially university students, who are interested in China. They also wanted me to provide background information to make the work accessible to the general reader. Therefore, while this is not an exegetical translation, I have tried to provide sufficient background information for the interested reader to delve further if so desired.

Aspects of How This Translation Was Made

Spelling and Usage

First, in accordance with the publisher's request, I use American spellings and usage (e.g., "plow" instead of "plough," the term "corn" instead of "maize"). In addition, historical context is also taken into account. For example, to avoid historical connotations of the word "peasant," *nongmin* ("peasant" or "farmer") is translated as "farmer" when talking about modern *nongmin* (i.e., those since the founding of the People's Republic of China).

Rhetorical Differences Between English and Chinese

Second, English and Chinese rhetorical traditions and devices are different. That means that logic as expressed in writing is also different; the results may be the same, but they are arrived at differently. For example, English writers normally construct a paragraph using an introductory sentence followed by supporting sentences. Chinese rhetoric sometimes inverts this, with the introductory sentence coming after the supporting sentences.

Even more common in Chinese is a relative lack of referencing (the use of pronouns to avoid repetitious use of antecedents) and substitution (the use of synonyms and paraphrase to avoid repetition). In English, referencing and substitution are marks of good writing; in Chinese, good writing is expressed in other ways, such as the appropriate use of classical allusions, proverbs, and elegant turns of phrases. For example, the four-character phrase *renduo dishao* (literally, "people are many, land is scarce") is used nine times in the book; by Chinese rhetorical standards, multiple uses of such a succinct and apt phrase that encapsulates the essential dilemma of Chinese agriculture are admirable. But in English, so many repetitions are not as felicitous; hence, the translation seeks to alter the phrase through referencing, substitution, paraphrasing, or elision.

Another example of this is that in the original Chinese version of this book, the binomial *nongye* (the standard translation of which is "agriculture" or "agricultural") is used almost 1,000 times and thus composes more than 2% of the text. This is normal in Chinese writing, but not in English. By the use of near-synonyms (e.g., agronomy, farming, crop farming, cultivation of the land, husbandry, tilling) and elision, this translation uses "agriculture" or "agricultural" far fewer times than the Chinese text uses *nongye*.

Translation Strategies: Domestication or Foreignization?

Finally, a decision had to be made to follow either a domestication or foreignization translation strategy. In this context, domestication strategy seeks to have the English translation closely conform to English and Western culture. On the other hand, foreignization seeks to extract as much information from the Chinese text as possible. The conundrum is that too much domestication gets in the way of English readers understanding the historical background and cultural and glimpsing foreign thinking and expression. On the other hand, foreignization, because of its exposures the reader to detailed explanations, exoticness (e.g., romanized foreign terms), and historical context, can overwhelm by taking the reader out of his comfort zone. Either "much is lost in translation" or the translation is difficult for those uninitiated in the foreign culture to understand easily. In general, I have favored a foreignization strategy.

For the sake of the specialist (and for those general readers who want an authentic and complete translation), this is essentially an unabridged translation of the original work.[14]

Important Background Terms and Information

Textual Notes

Part of the charm of reading a book translated from Chinese into English should be the ability of the English reader to access the Chinese world. Historical, cultural, political, philosophical, agricultural, linguistic, and even literary context need to be provided to the general reader to derive full sense and enjoyment from this book. Hence, there are a number of direct translations of Chinese figures of speech, footnotes for those who want a deeper understanding (but that can be ignored if one prefers a streamlined reading experience), appendixes, and a bibliography. Footnotes followed by "—trans." are those that I have provided; other footnotes are those of the author. Information in square brackets is mine; information in parentheses is the author's.

---

[14]This translation differs from the original Chinese publication in that updated statistics (to 2016) were provided to the translator by the author for inclusion in the first chapter and two sentences were elided for better readability, with consent of the author.—trans.

Appendices

To aid the reader, two appendices have been added, one for dynastic names and dates, and one for weights and measures. I am indebted to *Chinese History: a New Manual* for its detailed discussions of the periodization of Chinese history and of metrology. To provide further historical context, throughout the text I have added in parentheses dates for historical events, birth and death (when known) of personages, and reign dates for emperors. In addition, to save the reader from constantly consulting the appendix for weights and measures, I have sometimes made notes in the text on metrological matters or added equivalences in brackets.

Bibliography

The bibliography is enumerative, not descriptive. It contains books from the original Chinese bibliography, books mentioned in the text but not in the original bibliography (designated by a single asterisk), and books that I have added to the bibliography (designated by a double asterisk). References to classical works, with their myriad editions, are referenced in this translation (usually in footnotes) by the work's name, author, and chapter, so that they can be easily located, and hence are not included in the bibliography.

Two Important Romanized Terms

1. The first important term is *mu*,[15] the standard unit of land measurement in China, sometimes called a "Chinese acre." The choices for translating units of land measurement were to use: *mu*, acres, hectares, or square meters, etc. *Mu* is used throughout this translation because, when considering Chinese agriculture on its own terms, it really helps to think in terms of *mu*. In this book, productivity per unit of land, RMB value of output per unit of land, etc., are all referenced in terms of *mu*; to have converted *mu* to acres, hectares, or square meters would have unduly complicated and cluttered the translation. The reader needs only to keep in mind that there are approximately 6 modern *mu* per acre or 15 *mu* per hectare [i.e., the *mu* is equal to 667 m^2]. Incidentally, its dimensions have changed over time; the character *mu* is over 2,500 years old (though it is not attested in oracle bone inscriptions of the Shang dynasty).[16]

[15] Or *mou*, as it is sometimes pronounced and romanized.—trans.

[16] Some of the changes are noted in Appendix 2, Weights and Measures, and some are noted in the text of this book; for a fuller discussion, see Wilkinson, *Chinese History: a New Manual*, pages 557–558.—trans.

2. The other important term that is left romanized is *liangshi*. This is often translated as grain, food, foodstuffs, or staple foods (and is sometimes translated that way in this book, when it will cause little ambiguity). At other times, *liangshi* is used because it has no direct English equivalent and to use circumlocutions would prove awkward. As explained in Chap. 1, Section I, Subsection 1, *liangshi* is actually an important food category in Chinese that includes not only grains but also tubers and legumes. The actual Chinese word normally used for grain in the English generic sense is *guwu*. To give a sense of the relative prevalence and importance of the terms in the Chinese text, *liangshi* occurs more than 370 times, while *guwu* is used only twenty-three times.

There are other legal, political, or historical terms that are left romanized and hence foreignized, but explanations are given in the text or in notes. Historical terms include terms for land and tax reforms that played important roles in the evolution of Chinese agriculture and China's polity.

China's Land Tenure System

A basic knowledge of China's land tenure system is important to understanding its agriculture. It may not be familiar to many English readers and needs to be understood on its own terms. Chinese law lacks the concept of fee simple tenure (in which real estate belongs outright to individuals) that stems from English common law and other Western legal precedents. In China, no individuals actually "own" land as they do in America, say; they only have certain rights to it. As the author makes clear, because Chinese land tenure laws are still in flux, sometimes farmers' rights are not clear.

1. An important term for understanding contemporary land tenure rights is "household contracted land management" (*jiating chengbao jingying*), which has no English equivalent. This is a widespread system in China in which portions of land collectively owned by farmers are contracted out to individual households to be managed by them for set periods of time. This stems from the "household contract responsibility system (with remuneration linked to output)" that started in 1979.
2. "Land annexation" (*tudi jianbing*) is a term used in the book but with little explanation or context provided, since none is needed for educated Chinese readers. Suffice to say, in China's feudal society, annexation of land had long been the bane of peasants: The powerful, e.g., members of the aristocracy, corrupt officials, and major landlords have had a long tradition of coercing, bullying, or cheating the weak (e.g., peasants) out of land that the weak considered their own.

3. Today, in China rural land is sometimes taken by government through legal procedures that provide for at least some compensation to the owners of the rights to the land. Terms such as appropriate, arrogate, condemnation proceedings, confiscate, eminent domain, expropriate, public taking, requisition, seize, take over, etc., have specific meanings, sometimes with specific legal connotations. To avoid legal nuances, I generally use the term requisition or expropriate. The author discusses problems with the present system of rural land requisition and possible solutions.
4. Two other important phrases come up frequently that most Chinese readers will readily understand but that may be unfamiliar to the general English reader. The first is "reform of the countryside" or "rural reforms" (*nongcun gaige*), which refers to a panoply of Deng's reforms stemming from "reform and opening up," including decollectivization of the early 1980s and the subsequent move to a market economy in rural China. The second phrase is the policy of "reform and opening [to the outside world]" (*gaige kaifang*), one of the economic reforms termed "socialism with Chinese characteristics" started by Deng Xiaoping.

Problem Terms

A few problematic terms are dealt with in the text by using footnotes. One that needs to be mentioned here is *shidu guimo jingying* (literally, "appropriately scaled management"). It is found more than thirty times, primarily in Chap. 5. It could be translated as "appropriately sized," but it really means either "appropriately moderate-sized" or "appropriately large-sized," depending on one's viewpoint and context. It refers to expanding the size of small farms (which are defined as a farm with less land than the farm household can manage) into family farms (which are defined as farms that are predominately managed by family member that have enough land for the household to manage). In most cases, the basic idea is to increase the size of farm operations in order to increase efficiency and reap the benefits of economies of scale, so from an American standpoint, it might best be translated as "appropriately moderate-sized"; but from a Chinese standpoint, "appropriately large-sized" usually is the meaning.

Poetry

Penetrating views of China and Chinese agriculture are provided to English readers by the author's creative use of juxtaposing ancient poems alongside related modern topics under discussion. For the general Chinese reader, these poems elevate the tone of the book, add to enjoyment, and remind the Chinese reader of the importance of farming through the ages. For English readers, who may not be use to extensive use of poetry in books on agriculture, the use of the poems also drives home the importance of poetry in China and the love and respect that Chinese have for it.

I have translated the poems primarily for their meaning, without attempting to capture the rhythm or rhyme of the original. One should note that they are written in Classical Chinese and are frequently open to different semantic interpretations. The passage of time, sometimes thousands of years in the case of poems from the *Shijing*, has left modern readers guessing about the exact meaning of many lexical items; sometimes even more difficult to fathom is the thinking or mind-set of people from the distant past.[17]

The names of poems are given in Chinese *pinyin* romanization followed by their English names in parenthesis, by the poet's name, and then my translation; the poems are all easily accessible in numerous Chinese anthologies and so are not included in the bibliography.

Political Terminology

Finally, the translation can provide an entree into the thinking of Chinese officials, economists, agricultural specialists, and researchers. The English reader not used to modern Chinese political terminology may find such terms as "cadre," "comrade," "Plenary Sessions," (along with occasional boilerplate, jargon, hortatory language, and neologisms), etc., to be somewhat foreign. There are also long, complex sentences and copious information dealing with official meetings of the Chinese Communist Party and the enactment and changes of its laws and regulations. This is particularly true in Chap. 5, as the author gets down in the weeds to grapple with the complexities of rural land rights. For the specialist in land tenure law, it may be like Br'er Rabbit being thrown into the brier patch. For others, it can allow for new, trenchant observations about China's agriculture and a challenging introduction to Chinese political culture as it relates to it.

Style

In matters of style, in general I have followed the *The Chicago Manual of Style* (16*th Edition*) and *Chinese History: a New Manual*,[18] for sinological aspects.

[17]On occasion, one can clearly see this confusion in classical poems that have been translated into modern written Chinese: Different, experienced Chinese translators can come up with strikingly different interpretations of the same poem—indeed, sometimes translation of poetry boils down to making educated guesses.—trans.

[18]Endymion Wilkinson, *Chinese History: a New Manual*, pages xvii to xxiii.—trans.

Expression of thanks

My thanks go especially to Zhang Yunhua, the author, for his kindness in discussing with me the intricacies of Chinese land-use law and various other aspects of his book. Just as he comes across in this book has a genial and kind gentlemen, so he is in real life. It has been an honor to work with him. I also want to extend my sincere thanks to the School of English Studies at Shanghai International Studies University (SISU) for its support and encouragement. Furthermore, I am beholden to Chinese Culture Translation and Studies Support (CCTSS) in Beijing and its representative, Xu Donghao 徐冬皓, for helping organize and support this project. I am especially grateful for the help and encouragement of the following, all of whom are accomplished and talented translators, teachers, and leaders at their respective universities: Li ("Shannon") Mei 李梅, Sun Huijun 孙会軍, and Li ("May") Mei 李美 at SISU; Wu Yun 吴贇, Tongji University; Liu Yibin 刘翼斌, Zhejiang Gongshang University; and Meng Xingcan 蒙兴灿, Zhejiang International Studies University. I would particularly like to thank Zhang Chen 张琛, Renmin University of China, for his assiduous proofreading of *Insights into Chinese Agriculture*, and Helen Wang of the British Museum, for her help and inspiration. Finally, for their great help, assistance, and perseverance on this project, I would like to thank Qi Dongcheng 祁东城 and Zhang Lin 张琳, hardworking and conscientious editors at Shanghai Far East Publishers.

My apologies in advance go to the general reader for leaving the translation too foreignized, providing perhaps too much detail, and to the specialist for leaving it too domesticated by not delving deeper into agricultural, economic, historical, sinological, and other aspects. Nonetheless, it is my hope that all readers can garner useful insights and information from this translation. Any mistakes are entirely my own responsibility; I seek the indulgence and understanding of both the general reader and the specialist.

Shanghai, China
January 2018

Curtis Evans
Associate Professor, School of English Studies
Shanghai International Studies University

Contents

Prologue

Fascinating History of Agriculture

When you don't see the flower,
Its existence does not disturb your tranquility.
Only when you actually see it,
Are you suddenly touched by its beauty....

—excerpt from an untitled poem by Wang Yangming (1472–1529), noted Neo-Confucian philosopher of the Ming dynasty

Agriculture is the source of human civilization. Just imagine, if it did not exist then what would become of humanity?

If there were no farms to fill their stomachs, people would have to rely on hunting, fishing, and gathering wild fruits and nuts. Like animals, they would be dashing about all day trying to procure food. There would be no time to develop language, writing, or art, and there would be no way to carry out division of labor or mass production. The industrialized and informationalized[19] societies could never have come into being. Simply put, if there were no farms, there would be no civilization.

So, what is agriculture, after all? Let us start from the Neolithic Age.

[19] "Informationalized" or "informationalization" are translations of the Chinese government neologism *xinxihua* and are information age analogs of "industrialized" or "industrialization."—trans.

The Neolithic Age: Starting Point of Agriculture

Agriculture is the activity and the property used for raising animals and cultivating plants in order to produce farm products. It is man's conscious, regular, rhythmic exploitation of the growth of plants and animals in order to obtain food. In simple terms, it is that productive behavior of mankind that provides itself with food. In a broad sense, agriculture may be called "large agriculture." This means

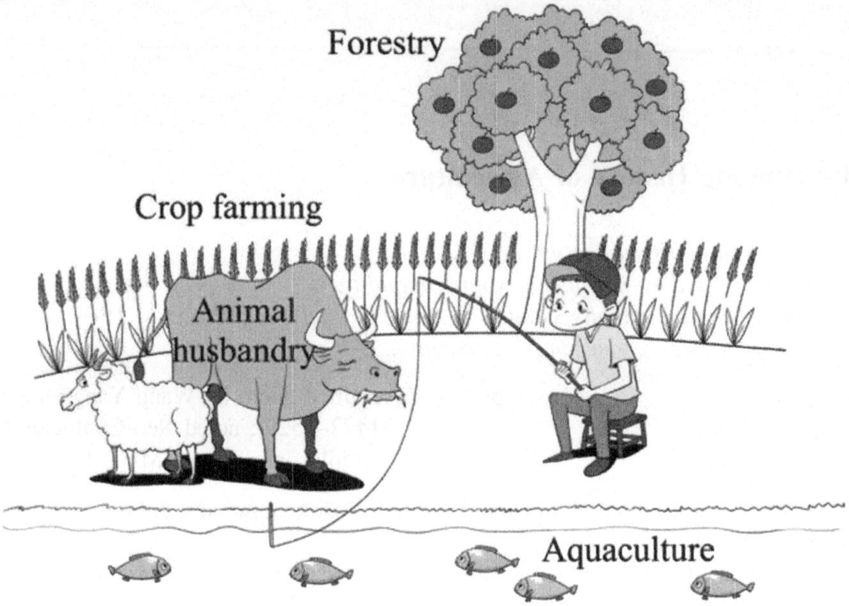

Agriculture is not confined to just growing crops

the crop farming industry, forestry, animal husbandry, aquaculture, and associated industries. In a narrow sense, agriculture can be called "small agriculture," when it merely points to the farming industry alone.

Was the rise of agriculture concomitant with the birth of mankind? It was absolutely not. In the process of more than a million years of evolution, the human race had passively depended on procuring food from nature. Updated to the Neolithic Age and equivalent to the advanced stage of primitive society in human history, about 10,000 years ago mankind started growing crops and raising animals, no longer passively dependent on the natural world to provide food. Instead, mankind became conscious of how to sow seeds in order to grow fruit and vegetables and how to domesticate wild animals to serve as a food supply.

Hunting was the primary method for primitive man to obtain food

The use of stone tools is a vital indication of the rise of agriculture. “Stone Age” is a special term in archeology that is also used in anthropology, history, and other disciplines. It is such a ubiquitous concept that a company has even developed a computer game by that name; a great number enjoys playing this game probably because they are thrilled and attracted by the primitive barbarism of that epoch.

If it were not for my attempt here to make clear the origins of agriculture, I would never have understood the concept of the “Stone Age,” which I first heard about during middle school. The Stone Age was an era when humans used implements made of stone and is one of the periods that archeologists use to delineate the early history of man. It can be divided into the Paleolithic and Neolithic ages, the Old and New Stone ages, based on the methods and levels of sophistication used to produce stone tools. The Paleolithic period, which stretched from two or three million years ago to a bit more than 10,000 years ago, featured mainly knapped stone tools. Starting about 10,000 years ago, with the time varying in different parts of the world, the Neolithic Age focused mainly on polished stone tools.

In both history museums and general museums, including the ones that I have visited (to name a few: the historical museums of Shaanxi, Zhejiang, and Fujian provinces), one can usually find displays introducing stone tools and archeological artifacts dating back to the Stone Age. By visiting these museums, one can sense how profound the influence of stone tools was on history. In these museums, one finds stone farming implements such as plows, axis, knives, shovels, sickles, and adzes. In the countryside today, one can see iron and steel farming implements still in use, the forms of which originated from those stone tools. Those stone implements all emerged from the Neolithic period and are important indications of this period. Of even greater import, they mark the origins of agriculture. It seems clear that its birth is the most vital and revolutionary feature of the Neolithic Age.

Let's take up stone knives and axes and fight!

The development of farming promoted the emergence of villages. According to the account found in *Renwen Shaanxi* (*Man and Culture in Shaanxi Province*),[20] by the middle of the Neolithic period, homo sapiens were capable of growing crops, domesticating and raising poultry, engaging in spinning and weaving, and stockpiling grain. In these ways, primitive farming helped shape the form of early settlements. The Late Neolithic period saw densely crowded villages on both banks of the Wei River that flows through the Central Plain in today's Shaanxi Province. Archeologists have discovered more than 30,000 sites in that area. The Banpo Neolithic archeological site east of Xi'an, the capital city of Shaanxi Province, dates back six or seven thousand years. Moreover, there is evidence of a primitive village reflecting a matriarchal clan society. The village, which lasted several hundred years, accommodated hundreds of people; farming was the primary long-term economic means that supported this village. Suffice to say, the formation and development of village settlements laid the foundations for the emergence of human civilization.

China was one of the origins of millet and rice. Carbonized grains (millet) and seeds of Chinese cabbage have been excavated from the Banpo site; they are examples of the earliest known specimens of these kinds of crops in China or anywhere else in the world. To this day, the people of Shaanxi and Shanxi provinces love to drink millet porridge. (In my childhood, I drank it twice a day as part of my breakfast and supper.) This must have been a custom established some six or seven thousand years ago that has continued right down to this day. Some remains of paddy rice have been discovered among the artifacts unearthed at the Hemudu

[20] Zhao Rong, *Renwen Shaanxi* (*Man and Culture in Shaanxi Province*), 2010.—trans.

Neolithic archeological site near Yuyao in Zhejiang Province (dating back some seven thousand years). This proves that in the lower reaches of the Yangtze River and nearby areas, China was the origin of the world's paddy rice cultivation. In recent years, carbonized grains and rachides of paddy rice that date back some nine thousand years have been discovered by the Zhejiang Cultural Relics and Archaeological Institute amid Yongkang City's Huxi ruins. Based on all these facts, research by both domestic and foreign scholars has indicated that China contains more early agricultural sites than any other place in the world. The main areas are the Yellow River basin and the Yangtze River basin; hence, these two are called "Mother Rivers" in China.

From Slash-and-burn Cultivation to the Use of Iron Plows Pulled by Oxen: The Evolution of Farming Methods

Slash-and-burn cultivation was a farming technique of the Neolithic Age; it is not only a synonym for the main farming technique of primitive agriculture, but also symbolizes it. Slash-and-burn is also called fire-fallow cultivation and involved using stone knives and axes to cut down a portion of a forest and then to burn the trees. This caused the soil to become soft and friable and to contain a large concentration of potassium and other elements as natural fertilizers derived from plant ash. Then, holes were dibbled in the ground and seeds planted. This method expanded the amount of land used for production of crops, but the yields were low. The next year a new area would be planted; hence, yet another name for slash-and-burn cultivation was migratory agriculture. In today's world, those using this method would be accused of causing man-made forest fires that destroy forest vegetation, would they not? From a contemporary point of view, slash-and-burn cultivation is an extremely primitive farming technique; but under the conditions back then, it was the best method for humans to start farming by adapting to and exploiting nature.

Turning over the soil is the first step in cultivation, so what were the earliest tools used to do this? A fork-shaped implement called a *lei* and a spade-shaped farm tool called a *si*, both handheld, were used in the Zhou dynasty's Spring and Autumn and Warring States periods just before the Qin dynasty. These were the earliest specialized tools for turning soil and are thus counted as the earliest agricultural implements.

The *si* is shaped something like a modern spade or shovel and was made of stone, bone, wood, etc. The difference between a *lei* and a *si* is that the tip of a *lei* is pointed while the two prongs of the *si* are flat. A bone *si* made from the scapula of a large cloven-hoofed animal has been excavated from the Hemudu archeological site. The *si*-style plow was produced in the mid-Neolithic era when agriculture was still in its primitive stage. Nevertheless, this method of cultivation was relatively advanced, making it possible for the primitive, nomadic way of life to move toward an agricultural civilization.

The *si*-style plow was not difficult to understand; even today, this implement is in use. Farmers still use iron spades to turn up soil to prepare ground to dibble in seeds in their own private gardens or small plots of land; these spades can all be counted as the descendants of the ancient *si* plow. When I was young, I helped my parents in our courtyard garden to grow vegetables; we adopted that same method. If we worked too much using this tool, we would quickly rub blisters on our hands, so work efficiency was low.

Farmers using an ancient *lei* implement to till the soil

The invention of the plow was certainly a great revolution in the history of cultivation. It enabled farmers to move from turning over only a small bit of soil and sowing seeds, to allowing a long row of soil to be turned over. This thus allowed the eventual transition from using a single farm implement to using a full set of farm implements. It marked an important advancement in agricultural civilization. Plows were initially made of stone; later, they were made of bronze and iron. Plows emerged in the Late Neolithic Age and led to the maturation and intensification of farming. The earliest plows were made of stone and pulled by manpower. In the Spring and Autumn and Warring States periods, the emergence of iron plows and the use of animal power to pull them were two of the most important technical advancements in the history of agriculture; these innovations lasted for several thousand years. Their use was premised on an intensified cropping system that depended on farming techniques such as crop rotation, multiple cropping, and intercropping. Individually managed farms came into existence in order to economize on labor, improve production efficiency, and broaden production scale. These then allowed the rise of slave-based farms that then in turn moved toward feudalism as the basis of agriculture. Cultivation using plows with iron shares and pulled by oxen is still practiced right up to the present in some areas. Twenty years ago in my home village in the Lüliang Mountains of Shanxi, we used this method to till our fields. Vivid in my memory of those years are scenes of adults plowing, while children followed behind playing. As a matter of fact, modern mechanized

Cultivation using an ox-drawn iron plow

cultivation that is used today emerged from ox-drawn iron plows; only now, the oxen have been replaced by "iron oxen," and the iron plow forms part of a set of farm implements that is bigger and more powerful than before.

Tales of the Fiery Emperor and Houji (God of Agriculture): The Dawn of Agricultural Civilization

Shennong [Divine Husbandman or God of the Farmers] was an alternative name of the mythological culture hero, the Fiery Emperor [Yandi or "Emperor Yan"], said to be one of the earliest ancestors of the Chinese people. The Fiery Emperor was surnamed Jiang and was a tribal or clan leader some 5,000 years ago[21] in the Valley of the Jiang River.[22] This tribe or clan was referred to as the Shennongshi or Jiangshi tribe. The Fiery Emperor's greatest contribution to farming was the invention of the plow. He taught people to use farm implements to sow the "five grains" (i.e., the most important "grains," including rice, wheat, and beans). He took his people, who were hunter-gatherers and migratory farmers, and transformed them into an agricultural society. In a classic tale that has been told over the last 5,000 years in China, Shennong was said to have tasted all kinds of flora and in this way he discovered herbal medicine. The Fiery Emperor was called Shennong, based on his ingenious contributions to agriculture; the name Shennong can roughly be translated as "Devine Husbandman" or "God of Farming."

[21]The chronology presented here should not be taken literally. Most traditional chronologies place Shennong and other of the earliest culture heroes much nearer in time to the Xia dynasty.—trans.
[22]I.e., the Qingjiang River, a tributary of the Wei River, in present-day Baoji, a county-level city in Shaanxi Province.—trans.

Among the descendants of the Fiery Emperor was a woman called Jiang Yuan; she was the mother of Houji, who is counted as the progenitor of the Zhou people and who established the Zhou dynasty; Houji's original name was Jiqi. In the time of the legendary sage kings Yao and Shun, he was put in charge of cultivation and attained great achievements. The people respectfully called him Houji, which means "King of Grains" or "Spirit of Grains." Houji became another of the important symbols of farming civilization.

His clan shared strong "farming genes" that were truly significant to agriculture. The Zhou people led by Houji became good at farming and attached great importance to it. Their agricultural foundation gave them the strength to defeat the Shang dynasty; the Zhou dynasty then formulated the political ideology in ancient China that made agriculture the basis of the economy.

In the oracle bone inscriptions of the Shang dynasty, etymologically speaking, the character *zhou* (周) is actually the character *tian* (田) with an added dot in each of its four squares. These dots appear to represent crops, making the ancient meaning of "zhou" to be "farmland."

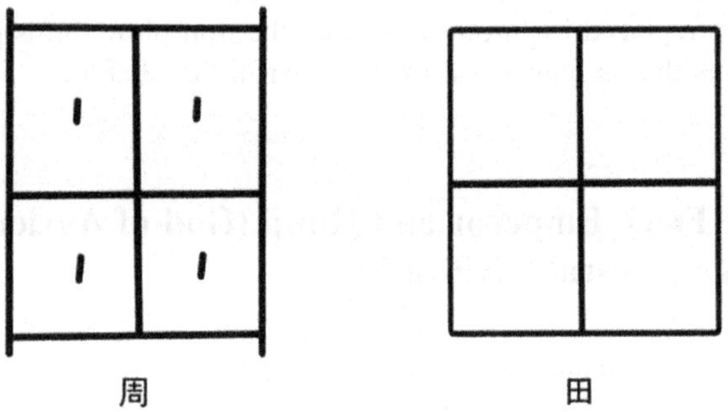

zhou and *tian* in the oracle bone inscriptions

Appreciation of Poems About Agriculture

Qiyue (*Seventh month*)

An excerpt from the *Binfeng* (*Poetic Airs of the Kingdom of Bin*) section of the *Shijing*[23]

[23] This is the longest poem found in the *Shijing* (variously translated as *Book of Songs*, *Book of Odes*, *Classic of Poetry*, etc.). It is the earliest collection of Chinese poetry (with 305 poems, all anonymous) and forms part of the classic canon of the literature and of Confucianism. The poems in it were composed over a period of five hundred years, from the early Western Zhou to the middle of the Spring and Autumn period. Because it is such an early work, the precise meanings of many of its poems have been lost and much is unclear [e.g., names of flora are often merely educated guesses, and (as in this poem) to what extent the poems reflect a slave society is a question].—trans.

.... The sixth month, we eat plums and grapes.

The seventh month, we cook mallow and legumes.

The eighth month, we knock down the fruit from the jujube trees.

The tenth month, we harvest rice;

With this, we make spring wine,

That we then use to help our bushy-browed elders to have a long life.

The seventh month, we eat melons.

The eighth month, we harvest the calabash gourds by snapping them off the vines.

The ninth month, we gather hemp seed,

Collect thistles, and chop down ailanthus trees for firewood.

All of this is done to feed us farmers.

The ninth month, we pound the garden areas around our houses to turn them into hard, smooth threshing grounds for the harvest.

The tenth month, we bring in the grain to the barn:

Both the early and late-sown millet,

Sheaves of rice, hemp, legumes, and wheat.

Alas, we poor farmers!

When the harvest is complete,

We then have to go to town and work on repairing the buildings of our sovereign.

In the daytime, we gather grass for thatching;

In the evenings, we twist some of it into rope to be used in roof repairs.

It is urgent that we get up on the roofs to repair them.

Then we again start the cycle of sowing all of the grain....

This poem describes work done in different seasons by the peasants of the Zhou dynasty and scenes from their lives.

The tribes of the Fiery Emperor and of the Yellow Emperor merged to form the embryonic stage of the present-day Chinese people. Historians consider the period of the Fiery Emperor and the Yellow Emperor to be the boundary between the primitive age and the period of civilization. It is now considered that the beginnings of civilization associated with these two great culture heroes mark the rise of agricultural civilization. Hence, this was also the beginning of Chinese civilization as a whole.

China has the largest population in the world. For long stretches of time (during the Tang, Song, and Ming dynasties), it was the world's most powerful nation. This greatness had a lot to do with China's farming culture. Intensive cultivation and rich production allowed the population to multiply and were the basis of national strength and prosperity. Throughout Chinese history, the chaos of large and small wars decreased its population and wrecked it politically and economically. Nevertheless, the Chinese people, relying on the incessant growth that is an inherent part of agriculture civilization, always eventually expanded their population, maintaining and strengthening their power.

Agriculture—An Age of Dominance

"The masses regard food as their prime want." Therefore, in order to be able to feed their people, the priority for generations of emperors and government rulers throughout history has been agriculture. Since the time of the Xia, Shang, and Zhou dynasties, for four thousand years in the annals of changing dynasties, the most important governmental imperative of monarchs was to ensure a flourishing agriculture that allowed the masses of people to eat their fill. When agriculture flourished, the country furnished; when agriculture waned, the country collapsed. Everyone is at least somewhat familiar with the fact that in successive dynasties, peasant uprisings nearly all corresponded to periods of agricultural dislocation and widespread famine. Other reasons were of secondary importance. In ancient China, the farming industry was crucial for China's economy; agriculture was the main part of the Chinese economy, and government revenue was dependent on taxes and levies on agriculture.

In the early period of slave society and feudal society, agriculture was of particular import. It deserved being called the primary industry. Both commerce and handicraft industries accounted for but a very small proportion of the economy. Only in the middle to late period of the feudal system did industry and commerce begin to gradually develop.

Owing to limited available data, one must more often than not use rough estimates for farm income. According to estimates of the scholar Liu Ti, in the 1600s, during the Ming dynasty, China's net agricultural output value was worth some 490 million taels of silver, and the gross domestic product (GDP) was some 900 million taels.[24] Therefore, the value of agricultural output accounted for 54.4% of GDP. In 1840, during the Qing dynasty, the net agricultural output value was worth some 3.07 billion taels of silver, the GDP was approximately 4.48 billion taels, and the value of agricultural output accounted for 68.5% of GDP. In accordance with the purchasing power parity (PPP) of the 1600s, China constituted about 25% of the world's total value of output, while by 1840 this percentage had fallen to only 20%. Thus, it can be seen that during that time, China's agriculture occupied an important place in the world economy.

In conjunction with the advancing development of modern industry and commerce, the relative value of farm production gradually decreased. With the founding of the country in 1949, the proportion of agricultural output value was less than one half of the value of the total economy. In the last two years, this figure has fallen to less than 10%. Nonetheless, this certainly does not mean that the importance of agriculture has diminished; just as before, it is still regarded as the basis of the nation's economy and of a stable society. Moreover, it serves as the basis for

[24]Liu Ti, *1600-1840 nian Zhongguo guonei shengchan zongzhi de gusuan* (*An estimate of China's gross national product from 1600 to 1840*), in *Jingji yanjiu* (Economic Research), No.10, 2009.—trans.

secondary and tertiary industries such as food processing, the food and beverage industry, the tourism and recreation industry, etc.

Taxes and levies on agriculture, primarily in the form of land taxes, were an important source of revenue for ancient Chinese governments. Notwithstanding this, in the wake of the development of commerce and industry, the proportion of such taxes and levies declined. During the Northern Song dynasty, 1077 agricultural taxes accounted for 30% of government revenue and those from industry and commerce constituted 70%. This shows clearly that the Song's agricultural society was moving toward an industrial and commercial society. However, because of changes in dynasties, progress was not unabated. By the reign of the Qianlong Emperor (r.1735–1799) of the Qing dynasty, such taxes and levies constituted 80% of revenues. By the Republican period, they still were one of the main sources of government revenues. After the establishment of the People's Republic of China in 1949, the proportion declined sharply: They constituted 14.7% in 1952 and, by the early 1980s, they were only around 2.5%. As a result, in 2006 the Chinese government officially abolished the agricultural land tax, which had been in existence in one form or another for some 2,600 years.

Precisely because of this, the guiding economic ideology of the imperial courts during China's feudal society was to pay attention to agriculture and inhibit commerce (including industry). That is to say, agriculture was the basic industry while manufacturing and commerce were secondary. However, it is difficult for modern man to understand why rulers of that time suppressed development of industry and commerce. Would not the development of commerce and industry have been a very good thing?

From today's point of view, industry and commerce are the biggest contributors to GDP, and it follows that most taxes and revenue also come from them. To explain the conundrum of why imperial China suppressed industry and commerce, one needs to consider the social and economic background of that period. In ancient China, agriculture was the foundation of the whole economy, so the rise and decline of agriculture related directly to the fate of the imperial court. At that time, the level of farm technology was low; there were no motorized machines and no advanced technologies such as scientific selection of good seed strains. In order to produce farm products sufficient to feed and clothe the people, to wage external war, and to secure sufficient land taxes and levies, it was necessary to rely on peasants to plant as much ground as possible. Ergo, state policy encouraged the people to engage in farming and at the same time tried to prevent peasants from leaving their farms to work or to engage in trade, since this was seen as having a negative impact on agriculture. Thus, the government put into practice policies limiting commerce.

Furthermore, the officials who governed state affairs at that time considered commerce (including handicrafts produced for sale) to be nothing but buying and selling, simply a kind of "skillful artifice" or exploitation of others by the clever use of money, instead of as a mean to create societal wealth. Therefore, the government disparaged and even suppressed industry and commerce.

Moreover, by binding peasants to the land and not letting them become displaced free agents with the freedom to seek opportunities elsewhere, they were easier for the rulers to control and thus to preserve national stability. Because this was an age of weapons of cold steel [in contrast to the later age of firearms], it was relatively easy for such displaced persons to take up arms and rebel. To rebel, all one had to do was gather a crowd and raise up swords and spears.

The earliest clear policy of favoring agriculture while disfavoring commerce can be found in the political reforms of Shang Yang (c.398–338 B.C.)[25] during the Warring States period of the Qin Kingdom.[26] Shang Yang's political reforms of the Qin State in 356 B.C. and 350 B.C. abolished the well-field system [also known as the *jing* system] and started the system of private ownership of land. Shang Yang levied feudal land taxes, called *chushuimu* ("first tax on acreage") in the Lu Kingdom (located in modern Shandong Province) and *chuzuhe* ("first tax on rice crops") in the Qin Kingdom. He also rewarded farmers for cultivation and for spinning and weaving, and encouraged the conversion of wastelands into productive farmland. Highly productive farmers were exempt from corvée labor duties; however, industry and commerce had restrictions imposed on them and were taxed heavily. Abandoning farming to engage in trade was forbidden. Using the system of favoring agriculture and disfavoring commerce as the basis for his economic reforms, Lord Shang played an active role in helping the Qin Kingdom to amass power. A robust farming system thus was the basis for the Qin Kingdom's unification under its rule of six other kingdoms of the Warring States period.[27] Unfortunately for Lord Shang, because his political reforms conflicted with the interests of the aristocracy, he was punished by dismemberment: his head, arm, and legs were attached to five horses that then pulled in separate directions. The very thought of it makes one shiver.

From the time of the first feudal dynasty when the Qin dynasty unified China under the first emperor of Qin, Qin Shihuangdi, until the very last of the feudal dynasties, the Qing dynasty, the policy of favoring agriculture and disfavoring commerce was the main strand that ran through state economic policies. It manifested itself in the following policies: the military colony system,[28] the equal-field system,[29] the government monopoly on selling land, and the system of government-run handicraft industries. All of these placed heavy taxes on commerce and industry, restricted merchants from taking up government posts, and interfered in their daily lives. The policy of favoring agriculture and disfavoring commerce strengthened the agricultural economy and thus overall state power during the early

---

[25]Known to history as Lord Shang, the great Legalist theoretician and statesman were noted for his draconian laws and strict and cruel execution of them.—trans.

[26]The Qin Kingdom was located in present-day central Shaanxi and eastern Gansu provinces.—trans.

[27]In 221 B.C. the Qin Kingdom unified China for the first time under an imperial system.—trans.

[28]This system, known as *tuntian*, consisted of stationing military garrisons on frontier areas in order for them to clear land, grow their own food, and be largely self-supporting. It originated in the Western Han dynasty.—trans.

[29]The equal-field system is also known in English as the land-equalization system. Its function is explained in Chap 3, Section II, Subsection 4.—trans.

period of the feudal age; at the same time, it consolidated imperial power. However, in the wake of the development of social productive forces, this policy started to obstruct economic development.

By the Tang and Song periods, strictures on commerce and industry had been somewhat attenuated; these two dynasties were also precisely those during which the development of the feudal economy was most magnificent. By the Ming and Qing dynasties, both a worldwide commodity economy and capitalism started to develop. Nevertheless, the rulers in China still clung to the policy of favoring agriculture and disfavoring commerce, which contravened economic laws. This resulted in China falling behind the rest of the world, because internally it did not vigorously develop its industry and commerce and externally it largely cut itself off from the outside world and severed trade relations with it.

No matter how brilliant, magnificent, honest, and incorruptible the emperors, generals, and ministers of the Ming and Qing dynasties are depicted in history books and novels, as long as they were not in step with the world's main trend toward economic development, they could only concentrate on minor matters to the neglect of major ones. They were thus penny-wise and pound-foolish. Without industry and manufacturing, there could be no construction of "armored ships and powerful artillery"; the result was that they could only submit to a thrashing in their clash with the Western powers. Not having the strength to defend the country and possessing such a huge market but refusing to engage in foreign trade, the results were that China was forced to open up trading ports after the two Opium Wars [in 1839–1842 and 1856–1860] and [make reparations and pay indemnities] after the incursion of the Eight-Nation Alliance.[30] In the face of foreign-made arms and artillery, the arrogant, feudal Celestial Kingdom could only bow its head in abasement.

[30]The Eight-Nation Alliance, made up of troops from Britain, the USA, Germany, France, Russia, Japan, Italy, and Austria, quashed the Boxer Rebellion in 1900.—trans.

Chapter 1
Does China Have Enough to Eat?

Farming is the basis of everything; it is what people rely on for life itself.
—Ban Gu and Ban Zhao, Hanshu (History of the [Former]. Han Dynasty) in Wendi ji (Chronicles of Emperor Wen)

1.1 Food Security: The Most Fundamental Domestic and National Affair

1.1.1 What Is Food Security?

In discussing a question, even more basic than logic are the concepts involved. If the concepts are not clear then the question cannot be delineated and the discussion will have no unifying basis on which to proceed. Concepts related to staple foods seem simple, but in fact they are frequently ambiguous; hence, for exploring the question of whether there is enough to eat or not, one cannot avoid starting with basic concepts.

Of what we eat every day, what can be counted as food crops? The basic Chinese term *liangshi*[1] can be divided into three large categories: grain, tubers, and legumes. Grain includes rice, wheat, corn, millet, buckwheat, oats, etc. Tubers include potatoes, sweet potatoes, and other such starchy root crops. Legumes include soybeans, mung beans, red beans, etc. In 2016, China produced 616 million tons of *liangshi*. The three largest amounts were corn, rice, and wheat (220 million tons, 207 million tons, and 129 million tons, respectively), for a total of 556 million tons. These made up 90% of the total production volume of *liangshi*, with tubers and legumes making up the rest.

The concept of food connotes a great many things. It includes *liangshi*, edible oils, vegetables, fruit, meat, eggs, milk, and sugar; all of these are edible farm products and foodstuffs. The rest of the world does not have the broad Chinese concept of *liangshi*, but only the more narrow concepts of food and grain. Both

[1]As noted in the Translator's Preface, *liangshi* is often translated into English as "grain."—trans.

Y. Zhang, *Insights into Chinese Agriculture*,
https://doi.org/10.1007/978-981-13-1050-8_1

domestically and internationally, there is no conceptual difference in the concept of "food." The Chinese concept of *liangshi* is close, but not identical, to the international concept of grain. The definition of grain used by the United Nations Food and Agriculture Organization (UNFAO) includes various kinds of rice, wheat [barley, rye, etc.], coarse grains [such as corn, sorghum, and millet], and various kinds of tubers and beans used as supplemental staples. By UNFAO's definition, soybeans are considered oilseeds instead of grains; this is a major difference with the Chinese definition, where soybeans are classified as part of *liangshi.*

In popular parlance, where there is food, there is food security; when there is nothing that can be eaten, there is food insecurity or food crisis. UNFAO's definition of food safety is: the guarantee that any person at any time is able to obtain sufficient food for their survival and health. Strictly speaking, what UNFAO is talking about is food security rather than the concept of *liangshi* security. The above concept came out of UNFAO's first World Food Conference in 1974. Afterwards, this concept underwent several redefinitions, and new connotations related to nutrition and food safety were introduced; but the basic connotation of food security did not change.

In China, there are two kinds of understandings of food security: one is the same as the international concept; this is the broad sense of the term "safety." The other kind is in the more narrow sense of *liangshi* security; that is, the secure supply and the balance between supply and demand of rice, wheat, corn, tubers, and beans. This latter concept is the one that the Chinese government uses and is the one generally understood in China; it is also the one used in this book.

Notwithstanding the above, there are also scholars who consider that a broader concept should be adopted, because in addition to eating *liangshi*, people also eat meat, eggs, milk, vegetables, and fruit, etc. Such scholars argue that adopting a narrow definition of food security could cause problems such as shortages of other kinds of food and difficulties in selling grains. However, to understand the question is to grasp the contradictions involved. The most important foods are those encompassed in the term *liangshi.* Meat, eggs, milk, etc., while also important, ultimately come from fodder and grain. Furthermore, this concept of *liangshi* security in its narrow sense has been used for many years in China and therefore it is appropriate to continue to use this original concept in this book.

1.1.2 "The Masses Regard Food as Their Prime Want"; a Country Lacking in Food Will Be in Chaos

What is more important than *liangshi*? Pearls are precious, but rice is king. Houses, automobiles, cell phones, and even silver and gold, power and status—none are as important as food. Food and drink are human's most basic physiological needs, and *liangshi* is the most important substance to satisfy those needs. Only after people satisfy their desire to eat and drink can they then pursue clothing, housing, and other such biological necessities. Certainly, this idea is consonant with American psychologist Abraham Maslow's (1908–1970) hierarchy of needs. These are the pursuits of higher-level safety (e.g., medical treatment for diseases and providing support for the elderly); demands for social interaction (e.g., making friends and courtship); and requirements for respect (e.g., positions and honor) and demands for self-fulfillment (e.g., giving expression to one's intrinsic worth and satisfying one's moral sense).

> The *Liji* (*Book of Rites*) says, "Food, drink, sex. Are these not the fundamental desires of people?"[2]
>
> "Master Gao said, 'Appetite and lust are part of the inherent nature of people.'"[3]

These two quotes can be interpreted to mean that food and drink are basic requirements; of the basic needs of life, *liangshi* is counted among the most important.

The concept of food security is generally used when referring to a county. The question is, do families also encounter the problem of food security? Of course they do. If the nation has a food security problem, then families certainly will be affected. If there is a shortage of *liangshi*, then it will be unavailable for sale or the price will be high. Furthermore, if a family's income is low and it is not able to buy sufficient food, then it must be subsidized or assisted in some way.

In the past, one of the most important reasons that each and every household in farming villages stored up grain was to prepare for famine and to thus guarantee their food security. In the 1980s, our family stored up corn in more than ten large earthenware vats that contained sufficient food for our family for one or two years. This was because my parents had suffered starvation when they were young. It was only in the beginning of the 1990s when I was in high school and following the reforms in the system of the purchase and sale of grain and the development of the market economy that the distribution of grain flourished. The staple food supply was then stable enough that my family no longer felt that it had to store up grain this way (This change, incidentally, helped to decrease the number of rats and mice.)

Appreciation of Poems about Agriculture

Guan yi mai (*Observations on scything of the wheat*)

— Bai Juyi (772-846) (also known as Po Chü-I), Tang dynasty poet

Farming families seldom have a slack month.
In the fifth month, people are doubly busy.
Last night the south wind blew,
Toppling over the yellow rows of wheat.
The womenfolk carry baskets of food,
In addition, children carry earthen jars of beverages
As they follow their mothers to the fields in sending lunch
To the able-bodied men at work on the Southern Ridge.

---

[2]*Liji, Liyun 9.xiii*. This passage is quoting Confucius.—trans.

[3]*Mengzi, Gaozi shang 6.iv*. Master Gao was a philosopher who sparred with Mencius over the question of man's basic nature.—trans.

Heat emanating from the hot earth
Scorches their feet and flaming sun sears their backs.
Their strength is exhausted, yet they do not feel the heat.
But the pitiless summer day is long.
Following behind them are the impoverished women,
Holding babies at their sides.
Their right hands glean broken ears of wheat,
Which they place in tattered baskets slung over their left shoulders.
I overhear the women exchange a few words,
And am overcome with sorrow:
"Our family's whole harvest has been handed over as tax;
Thus, we are gathering these remnants in order to allay our hunger."
Today what merit did I accumulate to deserve my wealth?
I have neither plowed nor raised silkworms.
An official's salary is some 300 piculs[4] *of grain per annum;*
At the end of the year, I still have grain to spare.
When I call this to mind, I am inwardly ashamed,
And for the whole day, I am not able to forget.

This poem depicts the busy time of harvesting wheat in the fifth month and the sorrow felt by the poet for the poor women who have lost their harvest to the land tax.

No matter how much science and technology have developed, nor how much society has progressed, historically food security is absolutely the most important thing and is basic to the governmental affairs of the rulers; there are no exceptions whatsoever to this principle, either domestically or internationally. What does it mean when the food supply is not secure? It means people must endure the torments of hunger! Do rulers have the heart to watch people starve to death? Of course they do not, because to ensure that everyone has something to eat is the basic business of government. How can rulers stand by and do nothing as they watch their people starve? Of course they cannot.

People starving to death is the thing that rulers can least bear, because the great turmoil caused by crop failures can lead to unstable political situations and even the collapse of governments. The Yellow Turban Uprising (184 A.D.) at the end of the Eastern Han dynasty, the peasant uprisings at the end of the Yuan dynasty, the Li Zicheng Uprising (1644) at the end of the Ming dynasty, and dozens of other peasant uprisings great and small throughout Chinese history, were all caused primarily by famine. The objective results were the weakening of national power, the rise of social unrest, and the promotion of dynastic change. Nonetheless, taking a broad historical view, so many instances of social unrest, peasant uprisings, and dynastic changes did not bring substantial change in the centralization of government power. The centralization of power, dominated by the imperial system since the Qin dynasty, lasted for more than 2,000 years, with a land system, a household registration system, and a tax system that were consistently fundamental; the only things that really changed were the surnames of the emperors.

[4]The *shi* is an old unit for dry measurement, particularly grain, equal to 133.33 lb.—trans.

"When grain is in one's hands, panic is not in one's heart."[5] To China, guaranteeing food security has been an everlasting task, one that can never be finished. Historical experience teaches us that once there is a great famine, having money is of no use. The problem of feeding 1.3 billion people is something that we must rely on ourselves to solve. In November, 2013, when convening a symposium at the Shandong Academy of Agricultural Sciences, General Secretary of the Communist Party Xi Jinping (1953–) used plain language to raise awareness and warn about food security: "Think of danger in times of safety." Even when today everyone can eat to sufficiency, the highest leader of the Party and country still issues this common saying as a warning, showing that it is of extreme importance that we are not remiss about the food security of this great nation of 1.3 billion people.

1.1.3 Famine Has Never Disappeared from Memory

Most of today's Chinese who were born before the 1970s endured hunger. Theirs was not the ordinary hunger of missing a few meals; rather it was the long-term hunger of not having enough to eat. Hunger was particularly etched in the memories of those who endured the Three-year Period of Difficulty.

Food rationing coupons are the best evidence of the scarcity of grain during that period. Those born before the 1970s should all vividly remember the grain coupons. In my memory, they were about the size of my thumb and were divided into those that were usable throughout the country and those that were only usable in local areas. Their face values were denoted by weight, in modern *jin* or *liang*. In my elementary and middle school period, every time we bought flour and steamed buns in town we paid for them with a combination of grain coupons and money. If one had money but no coupon, one could not buy grain or foodstuffs.

To relate another story, at the beginning of the 1980s when I was in elementary school, it was very rare for us to eat food made of wheat flour. On a bright sunny morning in 1982, for breakfast we had steamed scallion rolls made from wheat flour that were absolutely delicious. By midmorning, I longed to eat another roll; but fearing mother's reprimand, I simply filched one and rushed outside. Being extremely nervous, as I ran down the hill outside our courtyard gate I tripped and tumbled several times, right down to the bottom. Disregarding the pain, I immediately picked up the roll and continued to run. A neighbor saw this and yelled after me "Hey, kid, why're you running?" He could never guess that I just needed to find a quiet place to quickly eat up the roll.

Food ration coupons were used to purchase food during the period of the planned economy, when the government instituted a policy of state monopoly for purchase and sale of food. This system started in 1955 and officially ended in 1993, when grain markets opened up and edible grains and oils became available in ample

[5]This is from the first line of the well-known poem by Mao Zedong, *Shouzhong you liang* (*When Grain is in One's Hands*).—trans.

supply. During the period of the planned economy, the emergence of grain coupons was closely related to the problem of food security. These coupons were the most important manifestation of the period of the rationed economy, but there were also meat coupons, cloth coupons, etc. Grain coupons embodied the shortage of grain. Because the grain supply did not meet demand, the food security situation was grim. Depending simply on currency, it was difficult to adjust the balance between supply and demand. Therefore the government had to issue additional "grain currency" to bring back a balance and to safeguard market stability. After families in rural areas put into practice the system of household contracted land management, the supply of grain and other farm commodities gradually became abundant. By the middle of the 1980s, the use of food ration coupons became more flexible and their use declined until they were abolished.

Food ration coupons are remembered by a whole generation of people

中华人民共和国粮票 The People's Republic of China Grain Coupon

票样 Coupon Sample 拾市斤 10 *jin* 伍市斤 5 *jin* 贰拾伍克 25 grams

必须把粮食抓紧——毛泽东

"Hold on to grain tightly" — Mao Zedong

北京市油票 Beijing Edible-oil Coupon 油 oil

During the three difficult years from 1959 to 1961, China underwent a severe shortage of grain. Out of optimistic but erroneous judgment of economic conditions, insufficient knowledge of economic systems and economic laws, and under the inappropriate politicization of economic activities, in 1958 China inclined to the "left," taking advantage of governmental strength and power to carry out socialist construction known as the Great Leap Forward and the People's Communes. In the midst of the Great Leap Forward Movement, industrially the "Great Smelting of Iron and Steel" was central.

The target of that movement was to reach twice the production of the previous year by the end of 1958. The slogan was "Take Steel as the Key Link." Everyone was involved in the mining and smelting of iron and steel. Farmers had no time to grow crops, even to the extent of not taking time to harvest their produce such as grain, etc. Owing to the fact that raw materials were insufficient, even farm tools made of iron were gathered up and went toward making steel. In order to provide fuel for smelting, innumerable green hills were completely deforested, causing the ecology to suffer calamities that were difficult to recover from.

On the agricultural front, the central government, disregarding reality, rapidly pushed forward the movement of the Peoples' Communes. It advocated that the movement follow the principle of "First, Make It Large in Size" and "Second, Make It Collective in Nature" [I.e., let the People's Communes first of all be large in scale and second let them have a high degree of public ownership.]. But it fact, a better description would be, "First, Egalitarianism" and "Second, Free Goodies for Everyone." During the early period of the People's Communes, production was collective and meals were cooked in communal kitchens and served in communal dining rooms. As a result, responsibility and authority were not clear and it was difficult to supervise; farmers loafed on their jobs, and while it seemed that people were working long hours, in reality, efficiency was low. Communal dining also caused enormous food waste.

At the same time, boasting and exaggeration were in vogue. These revolved around such slogans as "The bolder the people are, the more productive the fields will be," and the widely disseminated false claims that new production records were being set for the highest yields. For example, the production from each *mu* was variously reported to be over 10,000 *jin* or even 100,000 *jin*. Based on such wildly exaggerated, false production reports, the nation taxed farmers for grain that in fact had never been produced. Nonetheless, theses taxes had to be paid in kind; this was the basic reason for why grain in farming areas was in such extremely short supply.

According to my mother's account, around 1960 in farming villages in Lüliang County grain production was in crisis. In addition, there was the "Big Unearthing of Surplus Grain" movement in which the farmers' large, subterranean vats of stored-up grain were dug up and carried off by government tax collectors. This was a big problem because too much grain was collected, which caused the farmers great difficulty in getting by. During that time, my mother and her brothers and sisters grubbed for wild herbs wherever they could and ate coarse grain [e.g., sorghum and millet]; they still did not have enough to eat. Father said that he did not go too hungry then, because he lived at his maternal grandmother's house. The party cadres in that village were honest and kind, and did not dig up all of the villagers' surplus grain.

1.1.4 Food Security's Big "Success"

"Farm crops are completely dependent on manure." This old saying means that growing crops well relies chiefly upon manure. People eat grain and then grain eats manure. There were no chemical fertilizers in ancient times, only "organic fertilizers." That is a common euphemism for farm manure that is produced by farming households and their animals. Its essential form of application is as a slurry made of excrement and urine from humans and farm animals. As an alternative, a slurry is sometimes made from chopped-up green plants and water, which is called "green manure."

Fertilizer is of great importance in farming; all farmers prize excrement. When I was small, I heard some grownups make a joke about so-and-so who was so stingy that every time he wanted to relieve himself he made sure to go on his own field, never on the land of others. The common expression, "Don't let one's own fertile water flow onto other people's fields" (a humorous euphemism for "Keep the goodies within the family"), comes from the same idea.

The earliest chemical fertilizer in China was imported in 1901 from Japan into Taiwan. Thereafter for fifty years, China had no chemical fertilizer production system of its own and so of course very little such fertilizer was used. Without chemical fertilizer, the grain yield per *mu* of hybrid strains, brought about by science and technology, would still linger at low levels. These strains would thus be unable to satisfy the food requirements of 1.3 billion people, a great many people would not have enough to eat, and the price of grain would be extremely high. According to my research colleague, Cui Xiaoli, in 1973 and 1974 China started to employ its relatively scarce foreign currency reserves to purchase complete sets of chemical fertilizer production equipment from America, the Netherlands, France, and Japan. This was the famous "43 Program," which referred to its total investment of $4.3 billion US dollars (at that time a huge sum of money) to import such equipment. At the Cangzhou fertilizer plant in Hebei, Mr. Cui noticed that even the refractory bricks that were used to build part of the plant were embossed with "USA" [despite the fact that China was perfectly able to manufacture refractory bricks and had been doing so for millennia]. At the end of the 1970s, farm villages were in the initial stages of reformation. A series of fertilizer plants started going into operation and in rapid succession the capability to produce and supply chemical fertilizer increased by a substantial amount. In the early 1980s, the increasing use of chemical fertilizer and the implementation of the system of contracted land management for farm families rapidly increased the yield of grain and other farm commodities. According to estimates of the Ministry of Agriculture and associated scholars, the use of chemical fertilizer has contributed about 50% to the grain yield of China.

1.2 Food Security—Expectations and Reality

It is not easy to explain clearly China's current and future supply and demand of grain and its food security situation. It requires synthesizing a large amount of data and research reports put out by government and scholars; it is a huge topic in itself.

Institutions and subordinate data analysis departments are all watching with close attention and concern and are analyzing the grain supply and demand situation. These bodies include: official governmental organs such as the National Development and Reform Commission, the Ministry of Agriculture, the Ministry of Commerce; subordinate departments that specialize in news analysis; a number of agricultural economists; and grain reserve entities such as the China Grain Reserves Corporation and the China Grain Group, etc. They also include resale companies, large international grain dealers, specialty consulting firms, and even financial corporations. There are a great many long-term, short-term, comprehensive, and classified analytical reports. To the best of my ability, I will attempt below to explain this situation as simply as possible.

1.2.1 Grain Is Now Essentially Sufficient

I would like to start with the demand for grain. What is the average annual consumption of grain per person? It is about 400 kg. Perhaps the reader may be wondering, "How could I possibly eat two pounds or more of grain per day?" This requires an explanation. One normally thinks of the consumption of rice or flour as what goes directly into one's mouth. Statistical tables show that with the rise in people's living standards, the consumption of meat, eggs, and milk increases as the consumption of foodstuffs diversifies; as a result, the direct consumption of grain is going down year-by-year. In 2015, the nationwide per capita direct grain consumption was 134.5 kg.

Nevertheless, excluding directly consumed grain rations, the per capita consumption of grain still includes animal feed that uses grain, grain for industrial use, and seed-grain for next year's planting, etc. When one eats meat and eggs and drinks milk, one is indirectly consuming grain that was used as animal feed; this still must be accounted for as part of per capita grain consumption. Grain is also consumed in producing processed food, and is required in making alcohol (for example, drinking a half kilogram of *baijiu*, the strong distilled liquor made from sorghum, is equivalent to the consumption of more than two pounds of grain). Finally, shrinkage [i.e., wastage], during storage and transport and in the consumption of food and drink, also has to be counted. Although the direct consumption of grain by urbanites is less than that consumed by rural residents, because the urbanite consumes more meat, eggs, and milk, their average consumption of grain (directly and indirectly) is not necessarily less than that of rural residents.

Let us move on to discuss the supply of grain. In 2013, China's grain production for the first time broke through the 600 million-ton mark, finally realizing the goal of "Ten Years of Consecutive Increases," and reaching historically high numbers. Compared to the 530 million tons in 2009 and 470 million tons in 2004, this represents an increase of 70 million tons and 130 million tons, respectively. Such huge growth reflects the capacity to enhance food production. It resulted from a combination of factors, such as state policy in the past few years, strong funding support, and favorable weather for crops during those years. In the years 2014 through 2016, grain production once more reached over 600 million tons per annum. As one can see from this, currently the supply and demand of grain is roughly in balance. So why have imports of grain shot up in recent years?

1.2.2 Grain Imports Have Been Massive

How much farm products have been imported into China and what is their proportion of total farm products in China? The answer is startling. China's supply of soybeans, vegetable oil, and sugar are all highly dependent on imports. The amount of imported grain in recent years has soared; the import of animal products has also expanded unceasingly. According to our general estimates, China's import of farm commodities has been equivalent to importing commodities grown on 700 million *mu* of farmland, an amount that would be equivalent to one third of the total cultivated land of China.

In 2016, China's import of soybeans totaled 83.913 million tons. The import of the Big Three cereal grains (wheat, paddy rice, and corn), in addition to barley and sorghum, totaled 21.795 million tons. Using China's grain yield of 616 million tons for 2016 as the basis for calculations, the quantity of imports accounted for 17% of the total. By using this measure, China's rate of food self-sufficiency does not even reach 83%! However, by international standards, which do not count soybeans as grain [but rather as oilseeds, yet which in China are counted as part of *liangshi*], China's rate of self-sufficiency in food exceeds 95%.

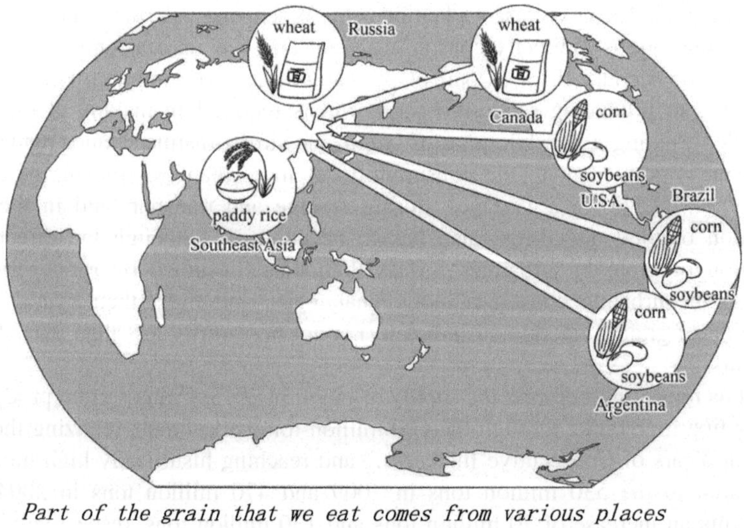

Part of the grain that we eat comes from various places around the world

The import of cereal grains has soared; the volume of imported corn is great. Prior to 2008 (with the exception of 2004), China's export of cereal grains (including wheat, corn, rice, and barley products) has exceeded its import volume. Nevertheless, after 2009, this trend has reversed and now the volume of cereal imports has started to consistently exceed exports and continues to go up. In 2016, imports of cereal grains reached 21.795 million tons. Among these, wheat, corn, and rice imports made up 3.412, 3.168, and 3.562 million tons, respectively. The imports of barley and sorghum totaled 5.005 million tons and 6.648 million tons, respectively. A total of 0.636 million tons of grain was exported. There was a net import of 21.361 million tons of grain, a decrease of 33.6% compared with the same period of the previous year.

Moreover, the import of distiller's dried grains with solubles (DDGS),[6] was 3.067 million tons. Imports of pork, beef, and mutton (the three main meats), that require grain and fodder to rise, have reached some 2.42 million tons, a great amount. This shows that although the gap in China's supply and demand of grain has been relatively small in recent years, it will be under pressure to continually expand in the future. To alleviate this problem, there is no alternative but to rely on imports. Furthermore, in the future, China's problem of tightening supplies of *liangshi* will still exist and imports of *liangshi* will continue to rise.

China is dependent on importing 86% of its soybeans, which represents more than 65% of the total world trade in soybeans. China is the world's largest soybean importer. In 2016 imports reached 83.913 million tons, but domestic production of soybeans amounted to a mere 13 million tons, more or less; 86% of China's soybeans came from foreign lands, making up more than 65% of the total world trade in soybeans. At the same time, imports of rapeseed totaled 3.566 million tons and vegetable oil imports reached 6.884 million tons, equivalent to the import of several tens of millions of tons of raw materials [i.e., rapeseed and other oil-bearing seeds].

The "High import" of agricultural commodities has been caused by short supply and topsy-turvy prices in China's farm sector. There are two reasons for this. One is that domestic supply is indeed insufficient; limits of water and soil resources and all kinds of other factors make China unable to satisfy the food demands of its 1.3 billion people. The other reason is domestic farm prices are higher than some of those on the international market. Cost, insurance, and freight (CFI) (including paying tariffs) of importing rice, wheat, corn, soybeans, cotton, and sugar are lower than prices of these items on China's domestic market. Ergo, the profitableness of importing farm commodities has led to sharp increases in imports. Over the last two years, the principal reason for the importation of large quantities of rice, wheat, and corn has been the low cost of imports, rather than problems with domestic supply — China currently has a huge amount of grain in storage.

In conjunction with this "High import," there is a "Low protection" of domestic agriculture. Under the framework of the World Trade Organization (WTO), China

[6]DDGS is the nutrient-rich by-product of dry-milled ethanol production that is made largely from corn products. —trans.

has substantially reduced tariffs on imported agricultural goods. In keeping with the Uruguay Round of negotiations of the General Agreement on Tariffs and Trade (GATT), tariffs on farm products dropped from a rate of 54% to only 15.2% in the decade after China entered WTO. Within the tariff quota of more than 20 million tons, a tariff of only 1% is collected on the three major grains (the tariff rises up to 60% for amounts exceeding this quota). Tariffs on the import of soybeans are a mere 3%, and there are no quota limitations.

1.2.3 From Now on, Supply Will Be Difficult to Provide

Forecasts of supply and demand for grain and farm commodities are also a difficult matter. This is not only because of the many factors that influence the output and demand of farm products, but it is also because of the substantial influence of different trade policies of foreign countries. There is no doubt that grain is crucial to China. Therefore, in forecasting supply and demand of farm commodities in China, scholars concentrate on grain. Nonetheless, among many factors that influence these forecasts, some are impossible to predict. Moreover, a number of indicators lack complete and systematic statistical data. Statistics and statistical approaches used are in discord. Added to this, forecasting methods are not the same. For all these reasons, domestic and international organizations and researchers come up with different forecasts for the supply and demand of China's staple foods and important farm commodities.

A relatively early and the most famous forecast was made by Lester R. Brown (1934–), director of America's Worldwatch Institute, who published the article, "*Who Will Feed China*?"[7] Brown argued that, in the wake of the increase in population and the improvement in consumption patterns, by 2030 China's food demands will increase by 85%, but owing to declines in productivity, reduction in arable lands due to urbanization, and environmental damage due to industrialization, by 2030 China's grain supply will decrease by 20% compared to 1994. When that time comes, if the annual per capita grain consumption is figured at 400 kilograms, then imported grain will reach 370 million tons; but global grain exports will not be more than 200 million or so tons. In that case, in 2030, China will only be able to satisfy 42.5% of its needs with domestic grain production; for the remaining 57.5%, it must depend on imported grain. By that time, China will not merely be unable to provide for itself, but the world will not be able to provide for China, either.

Brown's predictions caused a mighty uproar in China and startled the Chinese people. His predictions and analysis attracted a high level of attention from government officials and scholars. Afterwards, quite a few scholars published inquiries into or refutations of his predictions. Twenty years have passed since then. From

[7]Lester Brown, "Who Will Feed China?", *Worldwatch Magazine*, September, 1994.—trans.

today's point of view, the prediction made by Brown that China's grain requirements would increase to 640 million tons by 2030 was reasonable and it will soon come true. However, his prediction that total grain production would fall to 274 million tons on the basis of a reduction in arable land was too pessimistic; China at present has no problems producing over 500 million tons of grain.

However, Brown indeed brought up an extremely important and salient topic. Moreover, he was correct in analyzing the trend toward a shortfall in supply and demand. As predicted in the government's *Outline Plan for Medium and Long-term National Food Security* (2008–2020), by 2010 the average annual per capita grain consumption of Chinese residents will be 389 kg, which would require a total of 525 billion kilograms of grain. By 2020, the average requirement is predicted to be 395 kilograms per capita, requiring 572.5 billion kilograms of grain. My colleague, research fellow Xiao Junyan, has calculated that by 2020, based on a population of 1.43 billion people and an average consumption of 410–415 kg per person, the aggregate demand will reach 585–592 billion kilograms. At that time, China's domestic grain supply will experience a shortfall of 40–50 billion kilograms.

Other scholars have also made predictions, showing great differences. Speaking overall, the predictions of international organizations and specialists are more pessimistic than Chinese predictions. Many Chinese scholars think that China will be able to satisfy the food needs of its people mainly by exploiting its own resources. Judging from composites of various forecasts, future demands for grain consumption will remain over 600 million tons in the long term and, after 2020, it will perhaps reach 700 million tons. The tight supply of China's staple foods will persist; this is something that is difficult to change and so will require increases in the amount of imported food.

Owing to the government's high degree of attention to and support of the basic grain rations of rice and wheat, the rate of future self-sufficiency in these two grains should stabilize at above 95%, with imports not being too great. Owing to the continuous rise in meat consumption in China, the need for corn and fodder will steadily increase, and the import of corn and DDGS will grow significantly. If the amount of DDGS were converted into an equivalent amount of corn, then the quantity of imported corn would have already surpassed 10 million tons. It is even predicted that by 2020 the rate of China's self-sufficiency in corn will have fallen to around 70%.

One notices that the balance in the annual supply and demand of 600 million tons of grain is like a string that is stretched very taut. That is to say, with this so-called "tight balance" between supply and demand, there will barely be enough to eat. Food security can come true when it receives the highest priority of the Chinese government such that there is a steady increase in special supports such as continuous investments in and subsidies of grain production and when other farm commodities are imported at a high level. The "Ten Years of Consecutive Increase" is analogous to a small, cheap minivan that has been run two hundred miles at high speed; by exceeding its maximum safe speed, the vehicle has become unstable and is clearly at risk. Using current technology, room for growth of grain production is very limited. Furthermore, with respect to utilization of land, water resources, and chemical fertilizers, it has even gotten to the worrying point of "draining the pond to get at all the fish." To sum it all up, there is huge pressure on China's food security.

1.3 Food Security Under Huge Stress

In overview, China's food security outlook is grim and is under high pressure. I have summarized below the "Seven High Pressures" of food security.

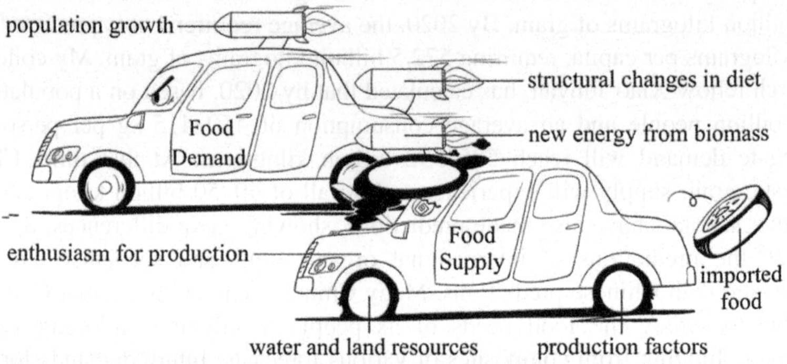

Food security under huge stress

1.3.1 Scarce Resources—Land and Water Are Insufficient

Chinese agricultural resources are in relatively short supply, forming a weak foundation for food production. China has too many people, there is too little land, and its water is deficient. The per capita arable land is less than 1.4 *mu*, equivalent to only 40% of the world's average. Another unfavorable aspect is that tillable land, including top-grade cultivatable land, is annually being reduced by more than a million *mu*. In 1996, China had 1.95 billion *mu* of cultivatable land, but by 2012, this had been reduced to 1.82 billion *mu*, a decline of 130 million *mu* during this sixteen-year period. Data from the State Council's Second National Land Survey shows that in 2009 arable land in China totaled 2.3 billion *mu*, exceeding by 200 million *mu* the data in 2009, which was based on the first survey. This does not mean that more arable land had actually been added, but rather it merely reflects that this new data is more accurate than that collected in the first land survey.

Chinese per capita and per *mu* water resources are merely 2,100 and 1,400 m^3, respectively, which only approaches 28 and 50% of world averages. As is known to all, China contains only 6% of the world's fresh water and 9% of the world's arable land, yet it has managed to solve the problem of feeding 21% of the world's population.

1.3.2 Large Population—"There Is not Enough Congee to Feed All the Monks"[8]

China's population is still increasing, urbanization is accelerating, and people's diets are improving. All these factors year after year push up the demand for grain. In 2016, China's population was 1.38 billion, an increase of 110 million people compared with the year 2000. The growth of population certainly brought about an increase in the consumption of grain. The percentage of people living in urban areas was up to 57.35, a 21.13% increase in comparison to the year 2000, and the population of towns and cities increased by 330 million people. Demand for grain was even larger because of the rich diets of urban people. Although their direct consumption of grain was less than that of people in rural areas, when one takes into account the amount of feed grain that is needed for producing meat, eggs, and milk, the grain consumption of urbanites exceeds that of rural dwellers. In fact, the aforementioned kinds of food require consuming relatively more grain; for example, it takes three pounds of grain to produce one pound of pork.

1.3.3 Poor Conditions—Infrastructure Is not Able to Keep Up

Although the country provides relatively strong support to improve farm production, irrigation and water conservancy infrastructure is weak. Thus, it is difficult to maintain stable yields and increase grain production. Approximately two thirds of China's arable land provides only low or medium yields. A few years ago when I was carrying out an inspection in Henan Province, I came to understand that farm conditions were relatively good, even in prefecture-level cities like Zhoukou and Zhumadian, where farming infrastructure was still very poor. But with the exception of the infrastructure in the agricultural program district, which was in relatively good shape, most of the infrastructure (ditches, irrigation canals, bridges, etc.) of other farmland there had been built ten or even twenty or thirty years ago; most of it was thus worn down, in disrepair, and decrepit after many years of use. Additionally, the construction of village-to-village road systems and expressways in recent years had wreaked havoc with the original irrigation and drainage systems. There had been insufficient investment in construction of irrigation and water conservancy projects, and standards had been lowered. All of this had clearly weakened the ability to irrigate the land during droughts and drain away water during floods. According to many cadre and common people, the bumper harvests of recent years have benefited from good weather conditions and the widespread use of good seed strains, with the former playing the more important role. But the

[8]This old Chinese saying means that there is not enough food to go around.—trans.

reality is that the foundation of agriculture there is fragile. The comrades in the Agricultural Development Office in Zhoukou suggested that we should not be too optimistic about the potential for great increases in production. The ability today to deal with droughts and drainage of waterlogged fields is hardly up to the standards of ten years ago. Once natural disasters occur, such as great floods or great droughts, they are bound to cause significant slippages in the production of grain. The popular belief is that in the past, 150 mm of rain did not cause disaster; now, even 50 mm of rain is disastrous.

In addition, there have been enormous negative influences on grain production due to the decline in the quality of arable land and the increase of severe pollution in China. This will be discussed in Chap. 2, Part I, "High Consumption" and "High Pollution" in Agriculture.

1.3.4 Low Income—Planting Crops Cannot Make Ends Meet

In recent years, the cost of agricultural production has climbed ever higher. The cost of raising crops has continually increased for farmers, but the increase in the selling price of farm products, in particularly that of grain, has not kept up. The low returns negatively influence the enthusiasm of farmers for planting crops. As statistics show, the three main grains provide an average yearly income of only about 250 RMB per *mu* (of sown area). A farming household, which on average has almost 10 *mu* under cultivation, would thus make a bit over 2,000 RMB, which equals the amount of money one could earn in only a month of hiring out as a manual laborer. After grain farmers assess their net returns on raising grain relative to net returns to be made by raising cash crops or laboring for others, they certainly will not have much zeal for planting grain. This problem is discussed in further detail in Chap. 4, Can Farmers Make Money?

1.3.5 Insufficient Rewards—Primary Grain-Growing Districts Suffer the Most

The provinces of Henan, Heilongjiang, Anhui, and other important grain-producing areas shoulder a great responsibility for providing adequate food for the nation. Nevertheless, agriculture is a disadvantaged industry. Main food-producing regions are also for the most part large crop-growing districts and financially poor districts; places where grain is an important part of district finances are usually financially straitened, with low average income for their people.

Since 2005, the finance bureaus of the central government and provinces have implemented policies rewarding counties that are large grain producers. To a certain degree, the government incentives eased the financial stress of these counties, and infused enthusiasm into rural governments. However, notwithstanding this, on average under these programs each such county annually received a bonus of only about 10 million RMB. Although these large grain-producing counties at the same time enjoyed general transfer payments, the amount per capita was very low. In the majority of counties, after spending money on providing some low-wage work [to the poorest farmers] and ensuring basic government expenses [needed to implement the programs], the actual usable financial resources were extremely limited.

There is a prominent disequilibrium between the interests of the main food-producing regions and the interests of the main food-distribution regions. The producing areas that are disadvantaged have strongly expressed their opinions.

Moreover, as part of the overall policy, those counties involved in these programs have had to assume all sorts of mandatory farm-related investment packages. The financial burdens have been so heavy that they have become millstones around their necks, inhibiting development of agriculture production. Some local cadres said that due to too many of these mandatory farm-related counterpart-funding support programs, they were faced with a dilemma: *On the one hand, they want to support agriculture and food production*; *on the other hand, owing to financial constraints, local governments can hardly carry out the counterpart-funding support programs.* As a result, the effects of these programs on the motivation of farmers and the development of agriculture and food production have not been as good as expected.

In 2008 while making a report on the grain situation in Henan Province, I brought up this problem. Under pleas from all sides, in 2008, the Third Plenary Session of the 17th Central Committee of the CPC [currently held every five years] mentioned the following in its *Central Committee of the Communist Party of China Resolution on Several Important Questions Regarding the Advancement of Reform and Development of the Countryside*: "The State will gradually abolish the financial contributions of counties and sub-counties to the public welfare construction programs, such as the reinforcement of endangered reservoirs and ecological construction specifically planned for Midwest China." With this, the financial problems due to supporting counterpart funded farm programs have been somewhat ameliorated.

1.3.6 Needs Are Great—People and Biofueled Cars Vie for Grain

Using yeast, corn can be processed into ethyl alcohol[9]; rapeseed and soybeans can be processed into biodiesel. Perhaps you did not know about this. The development of the biofuel industry has intensified the connection between the energy industry

[9]This can then be used to extend gasoline by making gasohol, usually a blend of 10% ethanol and 90% gasoline.—trans.

and agriculture. The forecast that farm products would be utilized for energy and thus that cars and people would compete for grain has become an ever-increasing reality. The rising prices of petroleum and other energy have brought about an increased demand for biomass energy sources. Using ethanol fuel as an example, there have been huge demands to greatly expand the planting of corn; this has infringed on the planting of cotton and wheat, which are grown in the same seasons [and on the same kind of land] as corn. As a result, this has led to the decrease in exports of wheat and cotton from some countries and has thus triggered gradual price rises for these two products on the international market. In the wake of this important change in the application of farm products, the way arable land is used in regions like South America, North America, Europe, and Russia has been transformed.

Because of the development of biofuels, the supply and demand of international agricultural markets has been remarkably altered: nations and regions where arable land is abundant rely less on international exports, while at the same time they have a disparately large impact on international markets. Countries that are major exporters of farm commodities in effect use the biofuel industry as a market regulatory device and formulate government policies that encourage their domestic businesses to develop biofuels and decrease agricultural exports, thus making it easier for them to manipulate supply, demand, and prices of international farm products. It also encourages them to maintain high international prices. In summation, using farm commodities as an energy resource is advantageous to countries and regions that are rich in agricultural resources, but for those that are deficient in such resources, it is very disadvantageous.

1.3.7 Dangerous Situation—Pressure on International Food Security Is also High

World population is increasing; as a result, urbanization in developing countries is increasing relentlessly and industry requires diverse agricultural products. These factors have brought about exuberant demand for farm commodities, causing increased pressure on their supply. Global demand for grain is continuously increasing at a rapid pace. Since 2005–2006, these demands have remained above 2 billion tons per year. In the past few years, the growth of biofuels has accelerated; the biofuel industry's demands on corn, soybeans, and other food staples have increased relentlessly. In summary, there has been acceleration in the growth of world demand for grain.

In December, 2013, data appeared from the Third Global Conference on Agriculture, Food and Nutrition Security, and Climate Change that was held in South Africa. It showed that over 800 million of the world's population [which was about 7.1 billion] could not be assured of food security. By 2050, the world's population may reach 9.6 billion people. When that happens, grain demand will be almost 70% more than current demand.[10]

Political and even military crises are something that the small number of socialist countries such as ours must consider. One simple worry is, in an unfavorable international situation, what if other countries try to strangle us? What if we cannot buy grain, even though we have money? This is one of the important reasons that China's grain reserves should be at a much higher rate than international standards (i.e., the recommendation that countries maintain reserves equal to 17 or 18% of their annual consumption). China's large population is another important reason to maintain a much higher grain reserve ratio than is internationally recommended. Even if no political or military risks emerge, once a regional or global food crisis occurs, major food exporting countries may reduce or even cease exports to foreign countries. For instance, in August, 2010, Russia temporarily halted grain and food exports due to a severe drought; during the 2008 global food crisis, Brazil and a few other countries also limited grain exports.

1.4 How to Guarantee Food Security

Guaranteeing food security is an important state issue; it is also an important issue for ordinary people. How can food security be guaranteed? National leaders are concerned about this. Numerous departments dealing with farming, land, water conservation, public finances, development and reform, and many other aspects of agriculture pay a lot of attention to it. Farmers expend great efforts on it, and ordinary citizens are also concerned. In a word, food security is truly something to be worried about.

Governments must attach importance to and support food security; this has always been the case with China. The Central Economic Work Conference that was convened at the end of 2013 recognized that to ensure grain security for the coming year was a top priority and hence thought in strategic terms about national grain security. That is, we must implement a national grain security strategy "… in which we take the initiative, that is based domestically, ensures productivity, imports grain in moderation, and is supported by science and technology. We must depend on ourselves for our own grain rations and gather domestic resources to concentrate on important areas. We must achieve basic self-sufficiency in food crops, and grain rations must be absolutely secure."

Following closely behind this was the convening of the Central Rural Work Conference, which restated this line of thought and emphasized that the Chinese people's rice bowls must at all times be firmly held in their own hands; our rice bowls should be filled primarily with Chinese grain. Only when a nation bases itself

---

[10]Ni Tao, "Dual Missions of Farming: Climate-Smart Agriculture," *People's Daily*, December 6, 2013.

on grain self-sufficiency can it control its food security; and only then can it control the development of its economy and society.

Hitherto, the Chinese government already introduced several plans. For example, there are the *Outline Plan for Medium and Long-term National Food Security* (2008–2020) and the *Plan to Newly Add 100 Billion* Jin [i.e., 50 billion kilograms] *to National Food Production* (2009–2020). The implementation of these plans and programs have had a positive effect in stimulating increased grain production; that there have been annual increases in grain production in the past few years is clear proof of this.

Ensuring food security is an ongoing battle, and one in which it is necessary to formulate strategy and tactics, attaching importance to strategy and remaining calm in implementing tactics. My viewpoint is that strategically, we must be self-sufficient in staple food grains and maintain overall balance. Tactically, we must provide all-out support and give prominence to key points.

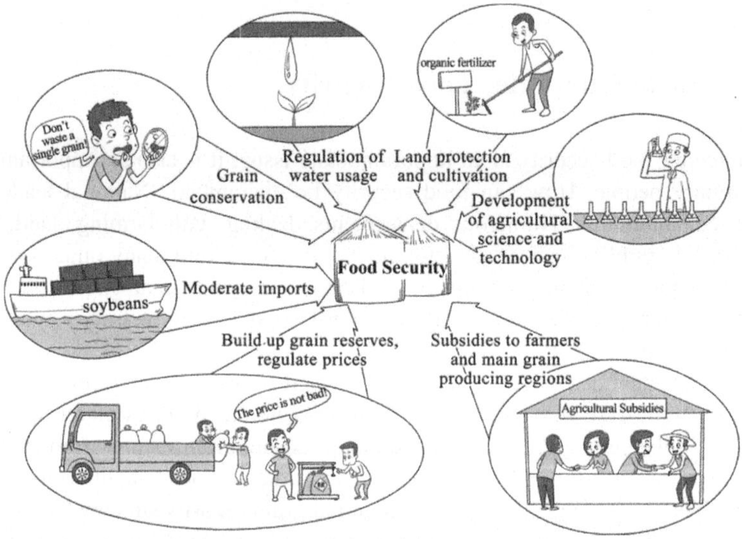

Strategies for Food Security

1.4.1 No Slackening on Self-sufficiency in Staple Grains; Overall Balance of Diverse Cereals and Grains

Self-sufficiency in staple grains, i.e., the strategy of establishing self-sufficiency in rice and wheat (the two most important grains), is now at about 95–100%; other grains reach overall balance through domestic production supplemented by imports from international markets.

First, there should be some changes in the statistical approach to measuring grain; I suggest leaving soybeans out of the category of *liangshi*. The *Outline Plan for Medium and Long-term National Food Security* (2008–2020) establishes that medium-to-long-range goals of self-sufficiency for *liangshi* should stabilize at 95% or more. For the subcategory of cereal grains [i.e., rice, wheat, millet, sorghum, and corn], the self-sufficiency goal is set at 100%. In light of the situation in recent years, these goals need to be revised. Soybeans, 80% of which are imported, are primarily used for oil extraction. Therefore, it is more reasonable that soybeans be categorized as oil-bearing seeds, the same approach used internationally. In so doing, we could not only integrate Chinese statistics on *liangshi* with international standards, but we could also be more clear and scientific in compiling statistics about self-sufficiency in *liangshi*. Otherwise, there is the great contradiction between the goal of 95% grain self-sufficiency and the reality that it is lower than 90% when soybeans are added to the equation. Without such an adjustment, there is no feasible way to make sense of these percentages.

Second, we must highlight rice and wheat as the two main staple grains and make our strategic target self-sufficiency and absolute security of staple grains. Yet, "grain rations" [i.e., *kouliang*] is hardly a strict definition; instead, it is an informal concept. As generally understood, grain rations are the grains that are directly put it into one's mouth; but *liangshi* that is fed to animals to produce meat, eggs, milk is not commonly understood to be grain rations. In that case, there is a large conceptual error here and a difference in specifications. The average urban resident consumes approximately 80 kg of grain rations per annum, while rural residents consume 160 kilograms. This does not square with China's annual per capita consumption of 400 kg of *liangshi*. From the perspective of supply, the annual per capita supply of 400 kg of *liangshi* is needed to ensure food security (i.e., grain or *liangshi* security). But the necessary annual per capita supply can easily be attained because actual grain rations make up less than half of the annual output. Therefore, it is more clear and precise to use the term "absolute security of staple food grains" [instead of "absolute security of *liangshi*"].

In a traditional sense, corn is one of the three staple grains. Because of its current principal use as an animal feed and as an industrial raw material, it does not serve as a staple grain as much as it did in the past. (En passant, I ate enough cornmeal to last a lifetime during my three years of junior high school). Furthermore, according to import trends and predictions, in the future China's self-sufficiency rate in corn may be only 70%. Can 70 or 80% really be counted as basic self-sufficiency?

Practically speaking, it is actually reasonable from a scientific point of view to formulate a strategy for establishing self-sufficiency in the two main staple grains, rice and wheat. In addition, we must achieve above 95% self-sufficiency in these two main grains by focusing our efforts. In so doing, we will place in our hands large bowls of rice and noodles made from grains grown by ourselves.

Finally, corn, tubers, and legumes need to be developed in overall balance. Corn is the "King of Animal Feeds"; the meat we eat depends to a large degree on corn. Feed corn occupies 200 million tons or 60% of the corn consumed in China. Corn is also an important raw material used to make starch and industrial alcohol. In the future, there will be even greater demand for corn. We must draw up a specialized developmental strategy for corn, based on our own efforts and thus free ourselves from imports.

In addition, tubers and coarse grains (such as sorghum, barley, oats, etc.) cannot be ignored; they are all important. Tubers such as potatoes and sweet potatoes are foods that everyone is fond of, and are important members of the *liangshi* family. Potatoes are categorized as both *liangshi* and vegetable, and almost nobody tires of eating them; even children who are picky eaters love to eat potatoes. Tubers in general have a high yield and are essential supplements during times of famine and crop failure. At the beginning of 2015, the Ministry of Agriculture initiated a strategy to include potatoes as an important part of staple foods. Buckwheat, naked oats (which the peoples of Shanxi Province and Inner Mongolia know how to cook and love to eat), millet, mung beans, and other miscellaneous grains and legumes are all important supplements to staple grains and are thus important for maintaining a balanced diet.

1.4.2 All-Out Support for Food Security; Taking Key Problems and Key Points into Consideration

Think this over carefully: ensuring food security is the main part of government's agricultural policy. If we look at those policies that aim to improve food production such as basic farm safeguards, protection of arable land, water conservancy projects, and comprehensive farm development, they all serve the main goal of ensuring food production. If we look at production, then agronomical breeding, plant protection, and mechanization all share this same main goal.

In addition, in order to safeguard the food supply and provide market stability, the "Four Agricultural Subsidies" (i.e., four categories of direct subsidies) have been implemented. They are the following: (1) subsidies for grain producing farmers; (2) subsidies to provide improved varieties of seed and livestock; (3) subsidies for rural farming materials [e.g., fertilizer, diesel, pesticides, agricultural plastic sheeting, etc.]; and (4), subsidies for procurement of farm implements and machinery. Other subsidy policies include those for price supports for grain, subsidies for the grain reserve system, and rewards to major grain-producing districts. The implementation of these government policies plays an extremely important role in ensuring food security. Of course, in order to accomplish this, there are relatively large costs in terms of manpower, materials, and financial resources.

All-out support for food security is popular in every country and China is no exception. In the past China did this, and now and in days to come, it will continue to adhere to the tactic of unconditional support for this goal. It involves all parts working together: agricultural production; production of *liangshi*; scientific and technological innovation and popularization; subsidies and rewards; establishment of food reserves; and price stabilization.

Based on such comprehensive support, it is necessary to give prominence to the following key points: One, strengthen infrastructure construction for irrigation and water conservancy. Two, strengthen protections and increase the intensity of support for arable land. Three, promote agronomy and the use of material and equipment. Four, continue to increase funding support for main grain producing areas. Five, refine government policies on grain subsidies and rewards and on price supports for grain. Six, optimize grain reserve systems and market control mechanisms. In addition, the last three key points, seven through nine, are to comprehensively improve capability and efficiency to ensure food security.

1.4.2.1 Key Point 1: Water Control

Water is the source of life. Since ancient times, water control for the use of agriculture has been a great task of all dynasties. At the dawn of civilization, the main task of Yao, Shun, and Yu[11] was to lead the common people in controlling water. "The Great Yu regulated the floods," is a famous ancient saying that has spread his good reputation throughout China. Li Bing [a famous hydraulic engineer who flourished c. 230 B.C.] and his sons constructed the great Dujiangyan Weir near present-day Chengdu in Sichuan Province, which has been irrigating the plains of Chengdu for some 2,200 years, rendering them a fertile and abundant "Celestial Storehouse," a cornucopia. Throughout history, how to control flooding of the Yellow, Yangtze, and Huai rivers has caused extreme worry and trouble to many emperors, generals, and ministers.

Today, in the wake of the progress of engineering technology, there is no comparison between modern times and the past in terms of preventing floods by water control; devastating floods are rarely seen in today's China. Irrigation networks are also relatively complete, and the ability to control and use water has enormously improved.

Henceforth, the key is to: consolidate large and middle-sized irrigation conservancy projects; positively develop small-scale irrigation and water conservancy facilities; and enlarge and intensify investment in basic construction of water resource facilities. The main targets should be the following: support of well-drilling and bridge and culvert construction; control of drainage and irrigation canals and ditches; and, the transformation of low-yield fields. Meanwhile, it is

[11]Yao, Shun, and Yu were legendary sage kings, paragons of regal virtue, and culture heroes of high antiquity at the dawn of the semi-mythical Xia dynasty. According to legend, Yao's reign was one of virtue and prosperity and he passed on his throne not to his son but to the worthiest man in the kingdom, Shun. Shun, like Yao, passed on his throne to the worthiest man in the realm, Yu. Yu was the legendary founder of the Xia dynasty. The three figured prominently in Confucianism, as it expressed state-sponsored morality and ethics. As seen here, they continue to exert strong influence in current Chinese thought and historiography.—trans.

necessary to: renovate farmland to high standards; enhance the ability of farmland to resist natural calamities; and steadily increase the yield of grain and the output of all crops. Furthermore, emphasis must be laid on reducing the waste of water, improving efficiency of water use, developing water-saving farming, and ameliorating problems of water pollution.

1.4.2.2 Key Point 2: Land Renovation

It is easy to figure out that if there were no cultivated land then there would be no production of food and that with a reduction in cultivated land there would be a corresponding decrease in grain. In short, with a drop in the quality of cultivated land, the amount of grain produced and its quality would decrease. Nevertheless, in the last thirty years, factory buildings and urban construction have occupied over 100 million *mu* of farmland. Fertilizer and an excessive use of pesticides and herbicides have impaired soil fertility. Wastewater, air pollution, and industrial waste products have polluted large tracts of land used for raising crops. Thus, both the quantity and quality of farmland are threatened. In the future, despite enormous pressures, great emphasis must be placed on ensuring that there is not a decline in either the quantity or quality of arable land. Land renovation is a feasible way to do this. By renovating and increasing the amount of farmland, we will be able to increase both its quality and quantity and thus enhance land-use efficiency.

1.4.2.3 Key Point 3: Scientific and Technological Innovation in Agriculture

In the more than thirty years since China's policy of reform and opening up, Chinese agronomy has developed to a high degree and now ranks at the forefront of the world. There have been all kinds of new scientific and technological applications and promotions of biotechnology and information technology in the realm of agronomy, which have been manifested under the rubric of "agricultural high technology." Its contribution to the rate of increase of farm output rose from 42.3% in 2002 to 56% in 2014. In 2013, its contribution to the rate of increase in production of *liangshi* reached 68%.

It is clear that science and technology are primary factors in the increase in farm production and that they are important dynamic forces in the development of agriculture. Innovations in agronomy have achieved abundant results; they have established a technological system for breeding high-efficiency plants such as super rice, genetically modified cotton that resists insects, sterile dwarf abortive wheat [which is useful in breeding new strains of hybrid wheat], and hybrid soybeans. In particular, there is leading international research on super rice and on new types of products through selective breeding that has developed large quantities of various high-quality, high-yielding, multiple-resistant, and new adaptive strains of crops suitable for broad areas. This has all led to the promotion of high-yielding and high-quality staple crops.

There has been important progress in technological research into zoological and botanical epidemic prevention and comprehensive insect pest control, thus strengthening the ability to prevent or control the spread of epidemic diseases in animals and plants. Wheat production has been completely mechanized from start to finish. Machine planting and harvesting of paddy rice, together with corn harvest mechanization, have been pushed forward at a high rate. Comprehensive farm mechanization has surpassed 60%, effectively replacing human and animal power.

Even though Chinese agronomy has been undergoing major development, there are still deficiencies. Chief among them are: the poor conditions of its basic science and technology; innovation is not strong; a few critical domains and important technologies are still behind advanced global levels; agronomy is out of synchronization with production applications; promotion of farm technology is lagging behind; and there are long-standing practices of "low application" of science and technology. Always an important force in agricultural development, in the future agronomy will continue to drive even more change and will require even more emphasis on: innovation, application, and popularization of science and technology; implementation of an innovation-driven strategy; and full support for the reform and development of agricultural science and technology.

First, accelerate the speed of the advance of agronomy: All-out support should go to scientific research that is both cutting-edge and on par with scientific research in industry. Establish world-class research organizations. Enhance innovation in these areas. Achieve major breakthroughs in key domains and core technologies, and what is more, help enterprises to become main centers of technological innovation. Encourage enterprises to integrate advanced, practical technology. Establish research centers for enterprise technology that are internationally competitive.

Second, intensify the transformation, demonstration, and promotion of scientific and technological achievements. Speed up the transformation of science and technology into real productivity. Strengthen the agricultural technique extension system through public welfare. Vigorously expand such government extension services.

Third, establish mechanisms for steadily increasing governmental investments in agronomy. Encourage social capital to invest in pertinent scientific research. Establish a diverse agronomical research investment system led by government, broadly involving enterprises and social organizations. Form stable and sustainable investment growth mechanisms.

Fourth, deepen the reform of the administrative management system of agronomic scientific research. Optimize scientific research's efficacy and organization. Strengthen the training of talented people engaged in agronomical research. Perfect the system for protecting intellectual property rights that involve agriculture. Enhance exchanges and cooperation with international agricultural research organizations.

Fifth, and finally, actively research, scientifically appraise, and prudently use technology for genetically modified (GM) agriculture.

1.4.2.4 Key Point 4: Subsidize Farmers to Produce Grain

Earnings are low for farmers who plant grain. The amount of money earned in a year from planting grain on 10 *mu* of land is less than what one could earn for a single month of performing manual labor in the employ of others. For this reason, direct subsidies to farmers who produce grain are still an important issue. A good many grassroots cadres suggest that subsidizing farmers several tens of RMB for each *mu* of land does not mean anything because the rate is just too low. They hold that direct subsidies are not as good as subsidies that concentrate on construction projects that can make a real difference. This viewpoint has some validity, but the "Four Subsidies," including direct subsidies for planting crops, are subsidies for the production of *liangshi*. These are given directly to farmers, and as such are direct incentives for food production.

Future work in this area should focus on the following: Perfect government policies on subsidies and increase the efficiency of subsidies. Implement subsidies for improved varieties of seeds of staple crops such as wheat, corn, and soybeans, under a generalized system of preferences so that numerous farmers can benefit from the policy. Expand the scale of subsidies for purchasing farm machinery. Increase the types of subsidies. Accelerate the process of agricultural mechanization. Give preference to subsidizing major grain producing farmers.

1.4.2.5 Key Point 5: Provide Incentives for Main Grain Producing Areas

Making a horse run and yet keeping him from grazing is not something that can be kept up long-term. The Central Government's Ministry of Finance still needs to do the following: increase direct transfer payments to grain-producing areas; assist in resolving the problem of insufficient financing of the main food-producing regions and counties; and give more support for social causes such as education, sanitation, culture, and transportation in grain-producing areas. In ensuring national food security, one crucial point is to increase the zeal of local governments of major food producing regions, just as the zeal of farmers to plant crops is motivated. In addition, government should take corresponding measures to encourage traditional farming districts to maximize their advantages and develop the grain industry in the framework of overall industrialization.

Furthermore, the government should: Provide additional preferential support in arranging and funding projects. Increase support for the agricultural by-products processing industry in the major food producing regions. Improve the ability of on-the-spot transformations. Increase the revenues of local governments and the incomes of farmers. Increase subsidies for grain production in main food-producing regions by providing direct government subsidies to those regions based on the quantity of marketable grain and by increasing the incentive funds for counties that produce large amounts of grain. Further, increase investments for expediting construction of systems for quality standardization in major grain-producing regions and for continuously improving the quality of staple foods. Establish as soon as possible mechanisms among financial departments at all levels and in each region in order to balance the interests among those who share responsibility for food security. Perfect the systems for funding food security and crop insurance. Reduce the percentage of funding costs that are borne by major food-producing regions. In addition, in areas and counties that produce large amounts of food, it is necessary to drastically reduce demands for farm-related local funding support for counterpart funds. Alleviate the already strained public financial burdens. Encourage the zeal among farmers in these regions to enthusiastically expand production of staple foods and general farm commodities.

1.4.2.6 Key Point 6: Appropriately Raise and Control Grain Prices

It is extremely difficult to regulate grain prices. If the price is low, farmers do not make money and will vote with their feet by leaving the farm or by not doing a good job, hence having a negative impact on food security. Since the price of grain is the basis for the price of a myriad of other items, if the price is high, it will induce the price of other goods to rise. Food prices, including the price of grain, is an important constituent of the consumer price index (CPI) and takes up more than one third of it. It is a weathercock for inflation.

Controlling inflation has always been essential for maintaining the health and stability of the macro economy. Therefore, it makes sense that controlling grain prices is an indispensable part of macroeconomic controls. Keeping the price of grain within a normal range is the goal of a country in order to maintain stable grain prices; the lower limit of the normal range of grain prices guarantees that farmers will not suffer losses. The government's current "weapon" to do this is the policy of maintaining grain price supports, i.e., a guaranteed minimum purchase price for grain. Every year, the National Development and Reform Commission drafts the guaranteed minimum purchase prices for rice, wheat, and other important types of staple foods. The head office of the China Grain Reserves Corporation is responsible for implementing a protected price floor for the purchase of grain to prevent farmers from being hurt by low prices. The upper limits of the normal range guarantees that grain prices will not rise to the point that consumers squeal with pain and that the CPI will not rise in an unseemly fashion. Therefore, when grain prices rise by a little, grain reserves will be released on the market in order to stabilize the price of grain.

It seems to me that the rising price of grain is constantly pushing up the price of everything else. As grain prices go up, so do the prices of oil, land, labor, and farming materials; in addition, the supply of cash also expands. This further pushes up the rate of inflation. But if there were no moderate increases in grains price, grain farmers could not survive, even when trying their utmost. Who then would raise the food? Farmers will be willing to plant crops only if the price of grain is reasonable.

Nevertheless, since the world has opened itself to international trade and "the earth is flat" [i.e., international barriers have fallen and global trade has become more equalized], the highs and lows in the price of grain are relative. Thanks to small-scale management and inefficient cultivation, China's grain prices have been relatively low, providing little profit to the farmer. However, in the past two years, Chinese grain prices have been relatively high compared to those on the international grain market. According to reports, prices in recent years have continued to be higher than those on international markets, and have climbed into the "highlands" of the world's grain prices. At the end of 2013, Chinese wheat and corn prices reached 2,500 RMB per ton, and were higher than American prices by about 1,200 RMB per ton. The price or rice was 2,900 RMB per ton, which were higher than the Vietnamese price by about 1,000 RMB per ton. In 2014, prices for Chinese farm products across the board were higher than those on the international market. The reason is because in America and in other countries that are large food producers, the business scale is large, the degree of mechanization is high, and the level of agricultural science and technology (especially GM technology) is high.

At present, grain prices in China are precarious. In comparison with other agricultural industries or with leaving the farm to do manual work, one cannot make money by planting crops. Prices should be moderately higher; but relative to the international market, Chinese food prices are already high. Ergo, the government will perhaps adjust its policy on grain price controls, turning the ever-increasing minimum grain purchase price into a target price subsidy. That is to say, when farmers sell grain, there will be a subsidy to cover the difference between the market price and the target price; low-income consumers would also receive subsidies to cover the price difference.

From the perspective of theoretical design, a target price subsidy system is quite good; but solving the operational problems inherent in such a system will be a hard nut to crack. Given different markets and different times of sale, how can one ascertain appropriate target prices or how can subsidies be provided to more than 200 million small-scale farming households? Perhaps it would be a waste of time, energy, and money, with unpredictable effects.

1.4.2.7 Key Point 7: Grain Reserves

As the old saying goes, "Storing up grain is like storing up gold"; if one has grain, there is nothing to worry about. Reserves are part of the foundation of stable grain markets and prices. Grain reserves are like grain prices: they also cause decision makers to worry. They have been a problem for millennia. When grain stocks are insufficient, panic ensues; when they are plentiful, numerous subsidies prove to be heavy burdens on government finances.

Furthermore, grain reserves squeeze demands on the populace and corporations. According to internationally recognized standards, the minimum sufficient grain reserves should be 17–18% of consumer demand. China's standard is higher than this level, probably much higher. According to media reports, from March to April in 2008 (at that time, rumors and predictions of a global food crisis were causing a clamor), Wen Jiabao, who at that time was Premier of the State Council, publicly declared a number of times that China's grain reserves were ample. There were 150 million to 200 million tons of reserves; the percentage of reserves was twice the world average. In the last few years China's grain production has increased year after year, reserve policies have been stable, and grain reserves have been sufficient to let everyone eat their fill with no worries. The current problem is that the policy of purchasing unlimited amounts of grain at the minimum price has resulted in the majority of the grain that farmers produce being sold to government granaries. As a result, there is no more room for storing the grain. Therefore, this causes a heavy financial burden on the government.

1.4.2.8 Key Point 8: Import Moderately

It is inevitable that China balances consumption demands by importing agricultural commodities. After all, we have large reserves of foreign currency to spend. If soybeans were not counted as *liangshi*, then the amount of *liangshi* that China imports would not be regarded as great. The annual amount of imported grains is 20 million tons. However, as previously mentioned, the imports of other farm commodities are certainly a lot. China should send a signal to the world of its requirements for stable, moderate imports of grain and other farm products. Utilizing international agricultural resources to reduce excessive use of domestic resources would be good for China. This would be conducive to China's agricultural rehabilitation and to sustainable development.

1.4.2.9 Key Point 9: Save Grain, Reduce Waste

According to our general estimates considering other data in the last few years, the amount of wasted grain (from harvest in the field, to storage, transportation, processing, and consumption) constitutes approximately 14% of total output. The amount of waste that occurs after it enters the consumer sector is astonishing. Some data show this waste to be about 40 billion *jin* [20 billion kilograms]. It is estimated that overall the amount of wasted grain is between more than 100 billion to 200 billion *jin* [50 to 100 billion kilograms]. Consequently, the whole society needs to conserve food and stop wasting it.

Who knows that every single grain in one's bowl is the result of so much bitter toil?[12]

As ordinary consumers, we should realize that a bowl of congee or of rice is not easy to obtain. In order that others may eat their fill, waste less. In order that we not suffer from "rich men's diseases" [i.e., costly, chronic diseases caused by overly-rich diets], remember to "Eat until only seventy-percent full."

---

[12]This is an excerpt from the well-known short poem, *Minnong* (*Pity the Peasant*) by Li Shen (772–846), a Tang dynasty poet.—trans.

Chapter 2
The Problem of Food Safety

> *The Way of Heaven provides benefits and does not harm. The Way of the Sagacious Man acts but does not contend.*
> —The Laozi (This is from the last two lines of the last chapter, Chap. 81. The *Laozi*, or *Daodejing* (*The Classic of the Way and Its Power*), was reputedly written by Laozi (c.500 B.C.), "The Old Master," Chinese philosopher said to be the founder of Taoism.)

Agricultural "High consumption" and "High pollution" harm both environment and people. Harm to the environment is an ecological safety problem; harm to people is a food safety problem (here it refers to the food safety issue involving the elementary production of primary farm products). Therefore, it is imperative to put into effect the strategies of sustainable development and food safety.

2.1 "High Consumption" and "High Pollution" in Agriculture

Agricultural production capacity has been enormously boosted by using synthetic chemical materials, such as chemical fertilizers, agrochemicals, and plastic sheeting, together with agriculture machinery and irrigation facilities. However, the unreasonable and excessive use of these external materials has brought about "High consumption" and "Low utilization" of energy and resources. At the same time, ecological and food safety issues have emerged, accompanied by "High pollution" and "Low safety."

Y. Zhang, *Insights into Chinese Agriculture*,
https://doi.org/10.1007/978-981-13-1050-8_2

2.1.1 *Excessive Use of Chemical Fertilizers and Agricultural Chemicals*

Just as people should not overeat, land should not be overfertilized, or there will be side effects. In 2013, the amount of chemical fertilizers used in China (converted to net weight) was 59.12 million tons, exceeding the rest of the world's total by 30%. In 1952, 1962, 1980, 1990, and 2000, the total usage of the rest of the world exceeded that of China by 758 times, 94 times, 4.7 times, 2.3 times, and 1.4 times, respectively.[1] The use of chemical fertilizers per *mu* in China is twice the internationally recognized safe upper limit of 15 kg per *mu*, but the average utilization rate is only about 40%. How does this excessive use of chemical fertilizer influence things? Soil acidification, salinization, and encrustation lead to declining fertility. Harmful chemicals such as heavy metals, nitrates, nitrogen, and phosphorus, etc., pollute soil, water, and even air. These ultimately destroy the ecological system and result in a drop in the quality of agricultural products.

The way that farm crops need chemicals is analogous to the way that people need medicine. By taking medicine suitable to the correct diagnosis and in appropriate amounts, diseases can be treated. Nevertheless, reckless or excessive use of medicine is harmful to health. In 2013, the amount of pesticides that China used reached 1.8 million tons. This was more than 1.5 times that used in 1990. Pesticides use was about two *jin* [i.e., 1 kg] per *mu*, but the utilization rate was only around 35%.

Apart from the excessive use of such chemicals, the danger is even greater from employing highly toxic pesticides that leave excessive levels of residues. Moreover, there is another common problem associated with unreasonable application of chemicals by farm households. In 2013, 2.49 million tons of agricultural plastic sheeting was used, more than four times the amount in 1990; each year hundreds of thousands of tons of such sheeting is left in the soil where it breaks down. Unutilized chemical fertilizers, pesticides, and residual plastic sheeting pollute the soil, groundwater, surface water, and air. This has resulted in varying degrees of pollution of over 100 million *mu* of land in China and has created ecological safety problems for both water and soil.

In addition to these problems, with increasing farm mechanization, China's agricultural energy consumption is gradually increasing. In 2013, 21.55 million tons of diesel fuel was used in agriculture; this was nearly double the amount used in 1995. Increased fuel consumption causes extra stress on the environment.

[1]See National Bureau of Statistics of the People's Republic of China, *China Statistical Yearbook*, 1981–2014.

2.1.2 Imprudent Use of Agricultural Irrigation

China is a country that has a serious shortage of water. Agriculture uses a relatively large amount of it. In recent years, agricultural has annually used approximately 62% of all water consumed in China, and yet China's use of such water, contrary to what one might expect, is comparatively slipshod. The method of flooding irrigation is ubiquitous, so water wastage is serious. The effective coefficient of utilization of China's farm irrigation is slightly over 0.5, much lower than the advanced world's level of 0.7–0.8. In the project "Evolution, Mechanism and Management of the Groundwater in North China," which is a National Basic Program (973)[2] of China, showed that in the last half century, the groundwater of the North China Plain has been overexploited by approximately 130 billion cubic meters. That would be enough water to form a groundwater depression cone equivalent to 70,000 km^2, which would rank as the largest in the world.

Approximately 70% of rivers in China are polluted to varying degrees. The waters of these polluted rivers are used to irrigate fields and nourish crops through irrigation systems. The heavy metals and harmful chemicals discharged by industry into water used for irrigation thus indirectly enter the food chain.

2.1.3 Serious Problem of Pollution from Livestock

The excrement produced by China's livestock industry causes another major source of pollution. According to statistics, the annual total amount of excrement produced by only hogs, cows, and chickens, the three main categories of China's livestock and poultry, is approximately 3 billion tons. A considerable amount of this animal waste, which could have been used as organic fertilizer, has been directly discharged into the environment without proper treatment, resulting in soil and water pollution. Another worrying food safety problem concerning livestock and poultry is the abusive and excessive use of veterinary medicines.

2.1.4 *Fifty Million* Mu *of Farmland Has Been Polluted*

According to soil surveys done by the State Environmental Protection Administration and the Ministry of Land and Resources, China has approximate 50 million *mu* of farmland that is heavily polluted. The amount of farmland that is

---

[2]Promulgated in March, 1997 (hence its moniker), its formal name is "*Guojia zhongdian jichu yanjiu fanzhan jihua*" ("*National development plan for key basic research*").—trans.

lightly polluted is probably even greater, perhaps over 100 million *mu*. In 2013, a certain province in the south of China experienced a "cadmium rice" scandal, which exposed the problem of heavy metals in soil.

2.2 Specific Food Safety Problems

When it comes to food safety, these are perhaps what housewives worry about most: pesticide residues in vegetables, veterinary medicine residues in meat, and heavy metal residues in grain. According to the classification of the World Health Organization (WHO), there are three kinds of dangers related to food safety: physical dangers from foreign objects in food, such as particles of rocks and strands of hair; chemical dangers, such as food additives, agrochemicals, veterinary medicine, and heavy metals; and dangers from microorganisms in foods, for example, disease-causing microbes. In producing primary farm products, the most serious problem for food safety is chemical danger. Its chief manifestations are residues left in food by pesticides, veterinary medicines, and heavy metals. However, in the sectors of food processing, market distribution, and sales, the dangers of food additives and microorganisms are more evident. Food safety problems come in all shapes and sizes and are too numerous to individually cite here.

Food safety problems are more and more serious

2.2.1 Ever-Lingering Problems of Agrochemical Residues

Every time you wash vegetables, do you first soak them for a long while, or when you eat an apple, do you peel it first? The problem of chemical residues in vegetables, fruit, and even grain has been vexing people. There is no doubt that the reasons for the abundant food on people's dining tables are due to methods and technology of growing plants that use large, modern greenhouses and employ appropriate equipment.

Another aspect also aggravates this problem. More than ten years ago as a doctoral student under the tutelage of Professor Kong Xiangzhi, I conducted an investigation on the use of agrochemicals among 353 farm households (from Shanxi, Shaanxi, and Shangdong provinces; these households were the valid samples obtained from a total of 833 sample households). Forty-eight percent had used pesticides that were extremely toxic. Among 1137 samples of farm chemicals, highly toxic chemicals such as phorate and omethoate[3] accounted for 29%. Moreover, in the crop-growing season they were used an average of two times. However, the Ministry of Agriculture at that time already explicitly forbade any use of these kinds of chemicals.

Although using the small sample data from ten years ago one is not able to extrapolate to the current situation, recent years have exposed quite a few problems concerning such agrochemicals. Exported vegetables are considered to already meet a higher standard than domestic vegetables, but every year there are still some instances of refunds paid to other countries following their investigations into excessive agrochemical residues in Chinese exports.

In 2011, we conducted a survey in a certain area. In a greenhouse that claimed to be environmentally harmless and that was recommended by local cadres and vegetable growers, several of my colleagues and I sampled some cucumbers and tomatoes without first rinsing them. After a few hours, two other colleagues and I started to suffer from diarrhea; one of them was so wretched that he developed a fever and had to be rehydrated with a saline transfusion. Although there were no tests at that time to confirm it, suffice to say, chemical residues cause most such misfortunes.

2.2.2 Ever-Intensifying Problems of Veterinary Medicine Residues

The problem of veterinary medicine residues is very serious. In order to accelerate the growth of livestock and poultry, there are hormones that are added to animal feed. In order to treat or prevent illnesses, excessive quantities or inappropriate use

[3]Both are insecticides and acaricides.—trans.

of antibiotics are administered. These all create the problem of veterinary medicine residues in farm animals, mainly manifesting in pork, chicken, duck, and fish. However, it is said that these residues can also be found in beef and even mutton.

Putting other such problems aside, let us talk about the "lean meat powder" [*shouroujing*] residues in pork. In China, pork is definitely the main kind of meat, constituting over 60% of all meat consumed. "Lean meat powder" is a generic name for a class of adrenal nerve stimulant pharmaceuticals such as ractopamine and clenbuterol [commonly called "bute"]. It is not used for treating swine diseases, but rather it is used to enhance protein generation and lean meat. Hogs feed "lean meat powder" have a higher proportion of lean meat and a better exterior appearance than those that are not feed it. However, residues of "lean meat powder" can harm people's nerves and hearts. At the beginning of the 21st century, the clenbutrol problem was already attracting attention from society. Government departments have taken a great deal of effort to keep watch on clenbutrol abuse. Nevertheless, right up to 2011, there was yet another clenbutrol incident that came to light involving a meat company in a province in Central China. The area affected was so large and the number of people affected was so great that it shocked the government and the public alike. It was almost as notorious as the 2008 melamine milk powder incident [explained in Preface 2 of this book].

2.2.3 Emerging Problem of Heavy Metal Contamination

In the past, it was only known that garlic chives could easily absorb heavy metals; therefore, in general people did not eat them. Nobody thought that grain also had such residues. In 2013, heavy metal residues were brought to everyone's attention following the "cadmium rice" problem that came to light in a certain area in the south of China. As a result, the problem of soil contaminated with heavy metals became known. Past data on soil contaminated by heavy metals has not been revealed, but absence of published data does not mean that there is no problem. Recently, the Ministry of Land and Resources and the Ministry of Environmental Protection have bravely made public the data that 50 million *mu* of farmland is moderately or heavily polluted. Let us give them a big thumbs-up in gratitude; before solving a problem, one first has to admit that there is a problem. According to plans of the National Development and Reform Commission and the Ministry of Land and Resources, there is a two-step solution to this problem that can be summarized as follows: Step one is that land unsuitable for cultivation has to be abandoned voluntarily by the farmers, with compensation paid by the government. Step two is to annually invest several tens of billions of RMB to carry out decontamination of farmland that is polluted with heavy metals.

2.3 What Causes Food Safety Problems?

Everyone detests the problem of food safety because it harms our health and causes all sorts of acute and chronic illnesses. Aside from detesting it, we should get to know its causes. Only by so doing can we apply the appropriate remedies to solve this persistent problem.

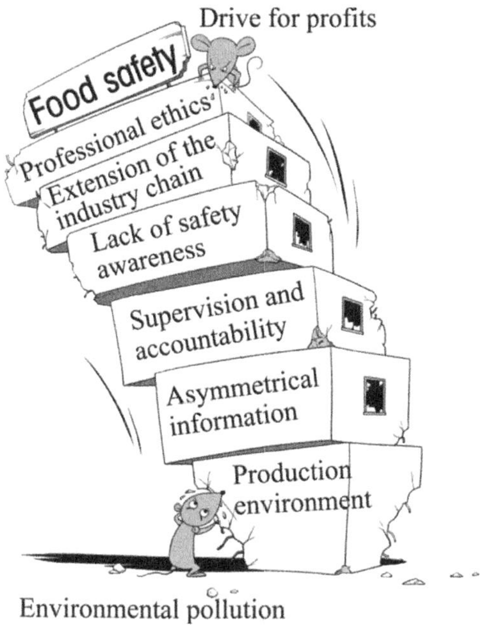

There are many causes of food safety problems

2.3.1 Improvements in Food Processing and Commercialization of Agricultural Products

Improvements in the commercialization of farm products have promoted the production of products that are no longer characterized by self-sufficiency but are oriented toward commercialization and specialization. Yet, along with the increased rate of commercialization, the inherent risks in processing and transportation are becoming greater than ever. Hidden dangers increase day by day. In the past, farm products were all sold as raw products, while today few are. In general, they have all undergone basic or intensive processing. Products are processed in order to, for example, increase storage life or enhance flavor. This influences the safety and quality of the food itself. During processing, additives and preservatives are used which to some degree jeopardize the safety of products. Food production is a high-risk business. From producing raw materials to intensive processing, from preservation to transportation, all facets of the process can give rise to safety problems.

Appreciation of Poems about Agriculture

Guo guren zhuang (Visiting the Farmstead of an Old Friend)

—by Meng Haoran (689–740), Tang dynasty poet

My old friend prepared for me the traditional welcoming meal of "chicken and millet" and invited me to come to his farmstead.

Verdant trees surround the village; outside its walls, the dark green mountains slant into the distance.

Opening the window, we face the threshing ground and courtyard.

Raising up cups of wine, we talk of "mulberries and hemp" [*i.e., life on the farm*].

By the time of the Double Ninth Festival [*held on the ninth day of the ninth month*], *it will again be the season to appreciate the chrysanthemum flowers.*

Compared to the self-sufficient agriculture in ancient times that is depicted in this poem, today there is a high degree of risk associated with improvements in modern food processing and commercialization.

2.3.2 Deteriorating Production Environment

As mentioned previously, the pollution of water, air, and soil is increasingly severe and cumulative and is an external environmental problem that harms the quality of farm products. As the external environment continues to deteriorate, some harmful substances continuously permeate into the environment where farm products grow and thus pose a very serious threat to their quality.

2.3.3 Consumers Lack Health Consciousness and Food Safety Knowledge

Dirty or ick,
It won't make you sick.

This widely known doggerel reflects the attitude toward food safety that old people in our farm village hold; this was even truer in the past. When I was young and begged for cake and snacks at my grandmother's, I always ate the moldy pieces first, saving the good ones to eat when they became moldy. This was also the sequence used by my mother, eating the stale food first and only then eating the fresh. More often than not, by the time the "fresh" was eaten, it had also become stale and perhaps had gone bad. In times when food was in short supply, eating one's fill was first priority; eating well came second. People's health consciousness and food safety knowledge were relatively deficient—fruit that had gone bad, flour that was moldy, or foods that were well past their prime, all went into our stomachs. In comparison with the past, has our knowledge progressed?

Over the last ten years or more, there have been some improvements, but they are still not enough. Salty vegetables contain large amounts of nitrites, but many people love eating them. Barbecued foods contain all sorts of carcinogens, but the roadside barbecue stands are thriving on summer days, emitting the intense smell of charcoal smoke and contributing their share to the thick haze of pollution. In addition, eating excessive amounts of salt puts serious strain on internal organs, triggering diseases like high blood pressure. The WHO recommends that a person's daily intake of salt should not exceed six grams; China's average intake level is more than twice this, and in the northern regions it is even more. I could go on and on like this.

2.3.4 *Economic Interest Drive Producers to Ignore Food Safety*

Farm households that produce raw agricultural products and food processing enterprises all consider their own economic interests as most important. For some of these producers, if highly poisonous pesticides are more effective and cheaper than pollution-free ones, then they will use them; if antibiotics are an effective way to treat illnesses, hormones can make chickens grow quickly, and feeding "lean meat powder" to hogs makes them easier to sell, then they will use them. After all, the residues from pesticides and veterinary medicines do not kill people immediately; as long as they are not detected, then there is no problem.

However, the producers themselves are careful to eat grain and vegetables that have not been sprayed with chemicals and eat livestock and poultry that have not been feed hormones. A popular story within the industry goes like this: A certain leader went to inspect a livestock breeder. He found that beside a large hog pen there was a small one that seemed to be different. He wondered what it was. The hog breeder said that the hogs in the large pen were fed chemicals to promote the growth of lean meat; those were for selling. Those in the small pen were not fed chemicals; those were for his family to eat.

This example does not represent the whole reality, but at a minimum, it clearly raises a question. Producers know about the problems of food safety but economic interests drive them to use the most effective methods to make money. Since the harm caused by the food they produce does not directly affect them, they choose to be indifferent to it. The money one makes is one's own; the harm caused by tainted food is someone else's problem. Is this not similar to the problem with pollution? Harmful food and pollution are both external to the producers, so in a situation where there is not enough supervision, naturally safeguards are ignored.

"Chickens that mature quickly" and "hogs that mature quickly"—of course some of the credit for this can be attributed to technological advancements in livestock breeding and raising. However, the problem of abuse of antibiotics and hormones still exists. Especially now under the model of intensive, large-scale selective breeding, hogs, chickens, and ducks from birth until slaughter are all enclosed in narrow pens or coops where they can eat but cannot move. They are fed feed that contains antibiotics and hormones and overdoses of medicines—even during the periods, when according to instruction labels, they are supposed to stop using them [e.g., in the weeks before slaughter]—in order to increase the survival rate of the animals and increase the production of meat. Is this, then, the outcome of modern agriculture? Are hogs greedy eaters or are people? In the anime production, *Sen to Chihiro no Kamikakushi* (*Spirited Away*) (directed by the Japanese anime master Miyazaki Hayao [1941–]), the mother and father of the young heroine, Chihiro, greedily eat until they turn into pigs. This shows a fine line of distinction between hogs and people.

"A few businesses only think of pursuing material benefits; the lack of morality and ethics is the key factor that gives rise to the problem of food safety." This was my view at the Fourth China Food Safety Forum in 2012. For the people working in the food industry, ethics is the critical issue at the highest level. The most bizarre food safety problems that have leaked out to the public are not about technology or regulations; they are about morality. The food safety supervision in China is one of the strictest in the world; with such strictness, why have we not done a better job? The reason is a lack of ethics, and this is not merely just related to food safety. Our forefathers advocated Confucian thought: in times of hardship, one should treasure oneself first; in times of success, one is expected to benefit others. But now, some people believe that when they are rich, they need not be benevolent; when poor, they can steal and cheat. According to some simple and even superstitious thoughts in ancient times, God is watching from above: good will be rewarded with good, evil will be requited with evil. Although this is not scientific, it is still useful. In the mind of some evil producers, what is morality? Can it be eaten as food? Doing bad things is okay, as long as they are not caught or even if they are caught out. Such people only care about making money and seeking profits. What do they care if they harm other people? Morality is something beyond economics, regulations, laws, or government policies. If everyone were virtuous, then food safety would not be a difficult problem and there would be no need for so many regulations and laws or so much supervision. In fact, morality is very simple: do not harm other people. This is what is called virtue.

2.3.5 Government Supervision Is Inadequate

Food safety belongs to the category of the commonweal and is thus one of the responsibilities of government. From what I have experienced since I started to get involved in research on food safety policy in 2004, I can say that the Chinese government takes food safety issues extremely seriously. Laws, regulations, and supervisory systems have gradually been adjusted, sorted out, and strengthened. Even so, the problem of food safety is indeed too complex, and there are still loopholes and weak links, especially at the grassroots level. Problems still exist, such as supervision not being up to standards and failure to exercise effective supervision and control.

In the realm of law, in 2006 and 2009 China promulgated different regulations: *Agricultural Product Quality Safety Law of the People's Republic of China*, *Food Safety Law of the People's Republic of China*, and *Regulation on the Management of Feeds and Feed Additives*, etc. These regulations have laid the legal basis for food safety supervision. In the lawmaking process, I participated in the work of submitting to the Development Research Center of the State Council of China suggestions and modifications for these laws.

What impressed me most was when, in April, 2005, under the instruction of researcher Han Jun,[4] Comrade Luo Dan and I made suggestions for revising the *Food Hygiene Law* (*Draft Revision*), a precursor to the *Food Safety Law*. Together we submitted twelve general suggestions and offered another fifty detailed suggestions. Among them was that the *Food Hygiene Law* be renamed the *Food Safety Law*, and that food safety risk analysis be added, both suggestions of which were adopted. From start to finish, the *Food Safety Law* took about six years to be officially introduced. During that time, there were numerous changes to the draft and many investigations, researches, meetings, and discussions were conducted. A great many comrades from the Legislative Affairs Office of the State Council, The National People's Congress, and some relevant departments, made arduous efforts to complete the law. Herewith, as a researcher involved with and as witness to this process, I offer a hearty thumbs-up of appreciation to those who devoted themselves with pains-taking care to this project. Recently, the *Food Safety Law* is being revised and public suggestions are being solicited, as required.

The supervisory system for food safety has gone through six major reforms and revisions since 2003.

The first revision was in 2003, with the establishment of the State Food and Drug Administration (SFDA). The SFDA positioned itself as an organization to coordinate and supervise food safety, but it was not as effective as expected.

The second time was in 2004, with the *State Council's Regulation on Strengthening Food Safety Work*. Further straightening out the responsibility of

---

[4]He is now the Deputy Director of the Office of the Central Leading Group on Financial and Economic Affairs and the Director of the Office of the Central Leading Group on Rural Work of the Central Committee of the Chinese Communist Party.

relevant departments under the principle that each department should be in charge of one certain sector, it took segmented supervision as its main method, supplemented by a variety of other types of supervision. It delineated food safety supervisory responsibilities. Departments dealing with agriculture, quality inspection, commerce and industry, health and hygiene, and food and drugs took charge of sectors concerning production of primary agricultural products, food processing, distribution, consumption, and comprehensive supervision, respectively. Among them, there was an obvious adjustment: shifting the supervisory responsibilities of food production and processing sectors from public health departments to quality inspection departments. However, at that time, when we made investigations and surveys of the supervision systems in the pork supply chain, we found that the quality inspection departments at the grassroots level were quite weak and did not match at all what the law called for. This was certainly related to the repeated mishaps that came out later in the food-processing realm in various counties and districts. This sort of multi-departmental segmented supervision system lasted until 2013.

The third revision of the supervisory laws for food safety was in 2007, with the establishment of the National Product Quality and Food Safety Committee.

The fourth was in 2008, when the SFDA was again incorporated into the Ministry of Health.

The fifth time was in 2010, with the establishment of the State Council Food Safety Commission.

The sixth and last revision was the establishment of the China Food and Drug Administration (CFDA), which integrated responsibilities for the following departments: Food Safety Office; Food and Drug Supervision Bureau; Quality Inspection Bureau's supervision of Food Safety in the manufacturing section; and the Industry and Commerce Bureau's Food Safety sector involving food safety in distribution. The Ministry of Agriculture is still responsible for supervision of the quality and safety of farm commodities.

The sixth adjustment finally unified supervision of primary agricultural production, food production, distribution, and consumption sectors. This suits China, as a country that produces large amounts of food and farm commodities.

It is worth mentioning that an adjustment shifted the supervision of on-site hog slaughter from the Ministry of Commerce to the Ministry of Agriculture. This should have been done earlier; for many years, livestock and poultry veterinarians, with supervisory responsibility all the time belonging to the Ministry of Commerce, did the primary level on-site disease inspection at abattoirs.

Looking back, I finished my doctoral dissertation, *Shipin anquan baozhang jizhi yanjiu* (*Research into safeguard mechanisms of food safety*) in 2005 and it was published in 2007. Based on a large number of cases and considerable empirical research, in its concluding chapter and the one on government policy implications, it put forward general thoughts about the reform of China's food safety supervision system. It suggested in the short-term to turn the multi-departmental system into a comprehensive departmental system. In the long term, there should be a centralized, single department to supervise all food products. Over the last ten years, these suggestions in my dissertation have essentially been in consonance with actual reforms that have since taken place.

In my chapter on government policy proposals, there were other suggestions: In the short-term, resolve problems that exist, as county-level public health departments and quality inspection departments sort out responsibility for food safety supervision and control functions. Quality inspection departments should strengthen food safety supervisory technology and take supervisory responsibility of the food safety sectors as soon as possible. It was also recommended that approval and supervisory responsibility over slaughterhouses that had resided with departments within the Ministry of Commerce be shifted to veterinarian departments that have actually been doing the work of onsite supervision.

It is clear that it is a long process from reaching consensus to actual reform.

2.3.6 *Failure to Distinguish Good and Bad Leads to Lemon Markets*

Does the food that you eat today have residues of pesticides or veterinary medicine, or is there harmful bacteria in it? The unaided eye naturally cannot see these things and they cannot be felt even when eating them; it is very difficult to distinguish food safety through human perception. For consumers, there is incomplete information. If the producer understands food safety information but the consumer does not, then between the two there is information asymmetry. With only incomplete and asymmetrical information, consumers need to be informed through specialized testing. Without this, consumers are in the dark. Under the circumstances of having unsound testing and authentication systems as well as incomplete and asymmetrical information, how can people be expected to wisely choose food and pay for it? When we cannot distinguish the good from the bad, will we pay high prices for so-called safe, high-quality food? There is no way. People roughly understand average levels of food quality on the market based on their past experience and pay the price accordingly. The result is that actual safe and high-quality foods do not sell well and have to be withdrawn from the marketplace, leading to markets being flooded with inferior, unsafe food products.

In economics, this phenomenon is known as "Bad money drives out good money." It has been observed throughout history, both at home and abroad. In ancient times, metal coins, etc., were circulated for use in commodity exchange. Take bronze coinage as an example: some of them were well made, for they contained high levels of copper of good purity; but some of them were just the opposite. In the markets, an interesting phenomenon emerged as people kept the "good coins" and circulated the "bad coins"; this is an example of "Bad money driving out good money." If the casting of the bronze coins was too good, and the copper in it exceeded the face value of the coins, people would even go to the extent of melting them down to sell the expensive copper in it for a profit. This led to currency shortages, market dislocations, and problems in banking. For example, there is a case mentioned in *Da Qing xiangguo* (*Prime ministers of the Great Qing dynasty*)[5] about Chen Tingjing (1639–1712), a high minister of the Qing dynasty. He once investigated a case of coins being melted illegally in order to sell their copper content. The law of "Bad money drives out good" is known as Gresham's law, after the Englishman Sir Thomas Gresham (1519–1579), who proposed it in the 16th century.

When "Bad money drives out good" is manifested in the market, a market for the sale of substandard products emerges. Economists call this a "lemon market," a term which was coined by the economist George Akerlof (1940–).[6] (By way of explanation, Americans commonly refer to articles of poor quality as "lemons.") After a certain period of time, "lemon markets" will be reflected by certain features in the social realm: Evil people with wicked behavior will be in power; virtuous people and good deeds will be hard to come by. If an elderly person falls down, will you help him up or not? When you run into a beggar, will you give him money or not? Will you be dependable and willing to work hard or will you be a shameless flatterer? Will profits be earned justly or by extortion? Similar questions can go on and on.

2.4 What Is to Be Done About Food Safety?

2.4.1 Lay Equal Stress on Supervision and Guidance

What is to be done when food safety problems continually pop up? It is necessary to implement a food safety strategy in which government and people in the food industry work together to do a good job. There is a great deal of research and discussion concerning the supervision of food safety, and it has many strategic mechanisms, with sub-systems, sub-departments, sub-industries, and sub-regions. The Research Department of Rural Economy of the Development Research Center of the State Council has also undertaken quite a number of surveys and studies and issued some reports and government policy proposals. To put it simply, to do a good job in ensuring food safety, it is necessary for both government and the food industry to conscientiously take responsibility. The government should enhance food safety inspection and service systems. The food industry should strengthen mechanisms to guarantee quality and safety. From the government's point of view, food safety is in the nature of a public service. Therefore, food safety supervision and service are important components of the national administrative system and

[5]Wang Yuewen, *Da Qing xiangguo* (*Prime minister of the Great Qing dynasty*), 2013. This is a historical novel.—trans.

[6]George Akerlof, "The Market for Lemons," *Quarterly Journal of Economics*, August, 1970.—trans.

within its administrative jurisdiction. Local governments are duty-bound to do a good job in this. Hence, food safety should be included as part of performance assessments of government departments, just as the GDP and work safety are. Only by doing so can food safety be assured.

This was the recommendation that we made three years ago when we did research into the success in Weihai [a prefecture-level city in Shandong Province], which was consistent with "strengthening local governmental statutory authority" that was mentioned in 2015 in the *No. 1 Central Document* of the Central Committee of CPC. For example, supervision of highly toxic and prohibited agrochemicals needs to be strengthened. Supervision of the production and distribution of such chemicals also needs to be strengthened in order to prevent those prohibited chemicals from entering the rural marketplace and thus becoming available for farmers to use. Enterprises which process, distribute, and sell food need even more strict supervision and should be brought under control in accordance with law. The government not merely provides supervision, but is also responsible for providing guidance in food safety; it must not only do a good job in supervision but also in technical training. Make sure that farmers and other producers are comfortable with the technology that is used to ensure the safety of farm products. For example, farmers should be familiar with technological standards concerning dosages of agrochemicals and the times when they should be used. Such educational work is an extremely effective way to encourage food safety for the public good.

People working in the food industry are primarily responsible for food safety. Whether farmers or enterprises, they should all abide by standards and regulations. Farmers must not recklessly apply pesticides and veterinary medicine, and corporations must not use illegal additives in food products. Corporations should monitor quality and safety of food commodities. They should do a good job in food quality and safety control in accordance with regulations and industry standards in the sectors of transportation and distribution, processing, and sales. At the same time, by agreements and cooperation, etc., those businesses in the supply chain of the food industry can ameliorate problems of safety throughout the food industry.

2.4.2 Develop Sustainable Agriculture

Farming belongs to the realm of nature; it depends on nature to follow natural laws. The Dao takes itself [i.e., Nature] as its own law.[7] Therefore, agriculture must respect laws of nature. This is imperative in order to transform development, carry out sustainable strategies, and return to the harmony of man with nature.

[7]This is a fundamental precept of philosophical Daoism quoted from the *Laozi*, Chap. 25. The full expression is, "People take Earth as their law; Earth takes Heaven as its law; Heaven takes Dao as its law; and Dao takes itself as its law." Dao can be roughly translated as "Nature".

Sustainable agriculture is an inherent demand of development and is the basis for food safety. Eco-agriculture, recycling agriculture, green agriculture, and low-carbon agriculture are all different names that represent sustainable agriculture from different perspectives. Hence, carrying out the strategy of sustainable agriculture is necessary to achieve agricultural modernity. This requires support from government from the following four aspects of financial investment and subsidies; industry policies; scientific and technological research; promotion; and preferential tax policies. In addition, all of the following four are needed to promote the reasonable and effective use of resources and energy:

First, implement the integration of ecological recycling and low-carbon farm technology; develop research into key technologies needed in ecological environmental control of agriculture; restore polluted farmland, eco-engineer farms and forests, and prevent or control disasters; and increase capabilities in agricultural resource conservation and ecological safeguards.

Second, enhance the efficacy of chemical fertilizer and agrochemicals and reduce energy consumption and environmental pollution. Accelerate the promotion of slow-release and controlled-release fertilizer technology and accelerate promotion of programs to provide free soil sample analysis to guide more efficient application of fertilizer. Guide and encourage farmers to utilize environmentally friendly technology. Increase investments in organic fertilizers. Use chemical fertilizers in moderation. Promote technology for prevention of biological threats to agriculture. Boost crop protection systems. Promote rational use of agrochemicals by farmers.

Third, enhance the ability to recycle by-products, such as livestock and poultry excrement and crop stubble, in order to turn waste into something of value.

Finally, increase water use efficiency and develop water-conserving irrigation. Promote more high-efficiency, water-saving irrigation technology (such as pipeline conveyance of irrigation water, sprinkle irrigation, drip irrigation, etc.), along with dryland farming, with its inherent water-saving technologies.

It is my wish that land everywhere may be pure.

Chapter 3
Rural Land System and Rights

The way of people is that those who have a steady occupation are steady of heart. Those without a steady occupation are not steady of heart. Without a steady heart, people are wanton and perpetrate crimes; there is nothing evil that they will not do.

—Mencius (Mengzi, Tengwengong, shang 1.iii)

Land is the mother of wealth, the most essential factor of production, the natural resource that is most typically in pressing demand, and the most important asset. Hence, a stable and effective land system with clear limitations and protections of land rights is the foundation of the economy and society.

3.1 One Land, One World

If there were no land, there would be no world. Understanding the land system is the only way to understand agriculture and the world—one land, one world; one land, one agriculture.

3.1.1 How Much Land Does China Have?

China's national landmass totals 9.6 million square kilometers, making it the third largest country in the world.[1] This is something that everybody memorizes in elementary school. Nevertheless, an additional important datum that not everyone is familiar with is that the surface area of the maritime area of China is more than 3 million square kilometers. A flower may have two branches; here I will only talk about one of its branches: China's landmass.

[1]For quick reference: 1 km^2 equals 100 hectares [or about 0.386 square miles or 247 acres], so 1 hectare equals 15 *mu* [or about 2.47 acres], and therefore 9.6 million square kilometers equals 960 million hectares [or about 3.7 million square miles] or 14.4 billion *mu* [or about 2.37 billion acres].

Y. Zhang, *Insights into Chinese Agriculture*,
https://doi.org/10.1007/978-981-13-1050-8_3

Of China's landmass, about two-thirds is composed of plateaus, mountains, and hills. Have you ever been to the Qinghai-Tibetan plateau, which is more than 3,000 m above sea level? The sky is clear, the grasslands boundless, the snow-capped mountains majestic, and the air thin. Have you ever walked among the lofty ridges and towering mountains that seem to extend forever? The mountains are side-by-side with streams that are side-by-side with mountains.

Back in the days before modern transportation and telecommunication, living among huge mountains meant being isolated from the rest of the world. Nowadays, benefiting from engineering technology to build bridges and tunnels, the network of railroads and highways extending in all directions has finally made it easy for mountain people to travel. Plains and basins are comparatively suitable for cultivation and for people to dwell; nevertheless, these sorts of areas only make up a third of China's land area. I would suggest that the reader look at a relief map of China, the kind with a rough and uneven surface that clearly shows elevations and valleys; this will help one to understand China's topography.

Of the 9.6 million square kilometers or 14.4 billion square *mu*, how much of it is arable land, forest, and other types? The Second Nationwide Land Survey data, which is the most complete and accurate so far, makes it clear that at the end of 2009 the extent of various types of land were as follows: arable land used to produce farm products such as grain, vegetables, and vegetable oil: 2.03 billion *mu*. This is the basis for China's food security and constitutes 14% of the total land area. Land used for the production of fruit, tea, etc., totals 220 million *mu* and constitutes 1.5%. Woodlands total 3.8 billion *mu*, which constitute 26.4%; this is more than that of arable land, and contains ecological forests such as pine and cypress and economically valuable forests such as those of walnut and jujube. Grasslands make up 4.3 million *mu* and constitute 30%; this is the single largest category of land. The boundless grasslands in summer on the Qinghai-Tibetan Plateau, in Xinjiang, and in Inner Mongolia present a landscape that makes one feel fresh and delightful. When fall and winter come, mutton from the grasslands is one of everyone's favorite table delicacies. The four types of land mentioned above are all agricultural land with a total of some 10.35 billion *mu* or about 72% of the landmass of China. In addition to these four types of land, there are others for the use of agriculture, including those for rearing livestock, for mechanized cultivation, for aquaculture, and for irrigation and water conservancy, as well as land used for thrashing and drying grain.

Finally, cities, towns, and villages, as well as industrial and mining areas, take up 430 million *mu*, or 3% of China's landmass. This amount is not great; however, over the last thirty years, expansion of cities and towns and development of industries have taken up over 100 million *mu* of prime farmland. Furthermore, land used for transportation takes up 120 million *mu*, or 0.8%. Bodies of water and water conservancy and irrigation facilities take up 640 million *mu*, or 4.4%. The rest is for other uses.

3.1.2 How Is Land to Be Managed?

The land system is the single most important economic and agricultural system and exerts an important and direct influence on market economy development, social justice, and citizens' property rights. This is the reason for the current national dedication to the reform and refinement of the land system. The land system is also perhaps China's most complicated economic system at present, much like a spider web. Its complexities lie in its design and content; in the internal complexity of financial interests of those involved; in how it influences other things; in its contradictory problems; in its reform processes; and in its future prospects. I do not think you can understand it by just looking at the vast ocean of books, articles, reports, and discussions regarding the land system. If you do not read them then you will certainly not understand; but if you do, the more you read the more confused you will become—it is not easy to make sense of this complex topic. Thus, I try in this chapter to make China's land system comprehensible by simplifying explanations and using charts as aids.

The land system involves numerous layers of complexity that contain many factors; if these are not dealt with properly, then it will be very difficult to give a clear explanation of what it all means.

From my perspective, it can be categorized into three systems: property rights, operational systems, and management systems. Who does a piece of land belong to and who has the right to use it? These are questions related to property rights, and the corresponding rules and regulations are the property rights system. What operational methods, forms of organization, and ways of interest allocation are applied to farm business operations? These are questions of operation, and the corresponding rules and regulations are the operational system. The implementation of concrete and detailed rules and regulations of the property rights system and the operational system constitutes the management system.

Simply put, according to this schema, in China's present land system the land property system means the state ownership of urban land and the collective ownership of rural land. That is to say, rural land is collective and used by farmers who possess rights of contracted management of the land, housesite rights, and usufruct.[2] The land operation system means the land of the cities and towns that are operated by government or enterprises. Rural land is operated on the basis of the household contracted management system together with a two-tier system that integrates centralized and decentralized management; this is also the basic operating system for farming villages. The land management system is both relatively complicated and concrete, and involves the protection of the quality and quantity of arable land, the approval of conversion of agricultural land to nonagricultural use, and the specifics of land rights. In the land system, the property rights system is the most basic. Once the main part of the land system of a nation is settled, then it is

[2]Usufruct (i.e., the legal right to use and enjoy the fruits or profits of something belonging to another) here includes the income and profit of collective lands.—trans.

relatively stable; very few difficult changes are required after that (e.g., the difficult transformation from private ownership to public ownership systems).

Nevertheless, the detailed contents of the property rights system can be altered (for example, the extent to which land property rights are entrusted to farmers). The operational control system that stems from the property rights system is relatively stable. Nevertheless, it may also be adjusted as the society and economy develop. The major changes brought about by the rural reforms in the early 1980s were in the land operation system, i.e., turning collective operations into a system of household contracted management. The new century has seen an increase of rural land transferred to this new system, after a large amount of the rural population drifted into the cities.

By the end of June, 2014, the area of arable land transferred to household contract management had already reached 380 million *mu*, constituting some 28.8% of the total arable land. There has also been a gradual increase in the number of different operational methods, such as cooperative operations, operations held in trust, and joint equity investment operations. There have also been some changes regarding the contents of operational systems. A relatively large leeway has been given to land management systems, especially in regard to some concrete management policies (Table 3.1).

Table 3.1 History of the development of China's land system

Era	Property rights system	Management system
Primitive society	System of clan members shared the land	Clan members held land in common
Slave society	State ownership of land	Land cultivated by slaves and commoners collectively; Later, cultivated by individual peasants
Feudal society	State ownership and private ownership coexisted; Most land was privately owned	Land cultivated by individual peasants; Or leased to tenants
Republic of China	Private ownership by landlords and farmers	Land cultivated by individual farmers; Or leased to tenants
The People's Re-public of China, to the mid 1950s	Private ownership by farmers	Land cultivated by farm households
From 1950s until the rural reform in the early 1980s	From private ownership by farmers to collective and state ownership system	From farm household management to unified, collective management
Since the rural reform in the early 1980s	Collective ownership by farmers	Farm household contracted management

3.2 Land System in Ancient Times

In primitive clan society, all clansmen shared the land and tilled the land together. After this stage of primitive society, the subsequent slave society practiced a system of state ownership of land that gradually evolved into a system of private ownership. In feudal society, a private land ownership system was the main component, with state ownership secondary. There was much more private farmland than public.

Before going further into an explanation of the land system in antiquity, land allotment systems have to be mentioned. In this context, "allot" [or "grant," "bestow," or "confer"] means that those at the top of society awarded land to those at the bottom, while "humbly receive" means that those at the bottom of society respectfully accepted something from those at the top. From the slave society to the feudal society, to varying degrees and in one form or another during successive dynasties, states carried out systems of allotting land.

These systems conferred state-owned land to both officials and common people. Some land was conferred in collective form (for example, the well-field system of the Western Zhou dynasty); later on, some was bestowed on a per household basis or based on the number of farmers in a household. Of these lands, some became private lands that would not need to be returned to the government and some had to be returned. Some of the land was confiscated from the previous dynasty and some was farmland that had been abandoned or wastelands that would have to be cleared. Some was granted to everyone, while some was granted to individuals based on their meritorious military duty, as acknowledgment of social rank, or as part of emoluments for an official position. Of course, the state would not confer all state-owned land, and always retained part of it to be used in other systems such as apportioning it to military garrisons for soldiers to grow their own food or allowing farmers to open up wasteland to grow grain.

Below I will focus on the well-field system, the *xiantian* system ("limited land tenure system"), the *zhantian* system ["occupied-field system"],[3] and the equal-field system. They were all land-allotment systems based on granting public land to farmers, but were distinctive in conception.

3.2.1 *Land in the Earliest Historical Periods Was State-Owned*

In the slave societies of the Xia, Shang, and Zhou periods, especially the Western Zhou period, a system of state ownership of land was put into practice. An indication that the Zhou kings possessed the land can be found in the couplet,

[3]The *zhantian* system was based on the size of land allotted to farming families, and the tax was called "tax on occupied land" (*zhantian ke*).—trans.

All the land under Heaven

Belongs to the King.

—Shijing (Book of Songs), excerpt from the Beishan poem

Based on a system of state ownership of land, the state awarded land to vassal lords, the aristocracy, government officials, and commoners, establishing land allocation, operation, and taxation systems. The Zhou kings enfeoffed [i.e., invested or bestowed] the feudal lords with estates of land and peasants. The feudal lords in turn bestowed parts of these estates to their high ministers, grandees, etc. They all could enjoy the use of the estates for generations, but could not transfer its lands by buying or selling. The Zhou kings retained ultimate title to the land. The feudal lords paid tribute [i.e., taxes] to the Zhou kings, while the peasants collectively cultivated land under the well-field system. In the early stages of the Zhou dynasty, the land system was principally represented by the well-field system.

The term "well-field system" was a figurative description. During the Western Zhou period, land was divided into many plots similar in shape to the Chinese character *jing* (井), meaning, "water well." [Hence, it is also known as the *jing* system.] According to what Mencius wrote, recorded in the *Mengzi* (*Tengwengong, shang 1.ixx*), "Each square *li* [i.e., approximately a third of an English mile] is divided into a matrix in the form of the character *jing*, constituting a total of 900 *mu*. Eight families individually farm the outer eight plots of land that are composed of 100 *mu* each; the middle plot of 100 *mu* belongs to the state. Together they jointly tend this central field. Only after completing their communal work do they dare tend to their own affairs [i.e., work on their own private plots of land]."

One hundred *mu* at that time would be about thirty or so of today's *mu*, which would have been sufficient to supply the needs of a family.

Farm work on the central, communal plot constituted corvée labor; the crops grown on the communal plot were handed over to the state and so were equivalent to a tax paid by corvée.

3.2.2 *Privately Owned Farmland Emerged in the Spring and Autumn Period*

After the Spring and Autumn period, labor productivity greatly improved when farm implements made of iron were introduced. Peasants took more interest in laboring on their private fields; therefore, the communal plots went uncultivated. After large amounts of wasteland were brought under cultivation, these newly cleared lands were kept in private hands, effectively turning them into private property. Simultaneously, by means of conveyance, robbing each other, and by bestowal from the state, there was a sharp growth in private land held by the aristocracy. As a result, the original equal-field system, which depended on the income from communal fields, faced financial difficulties.

The Zhou kings and their vassal states, in order to preserve their normal financial revenues, could no longer avoid reforming the land and tax systems. Therefore, they put into practice land reforms and changed the well-field system, which had allocated eight families to farm eight plots of land, with a central plot to be communally farmed as a form of corvée (known as *zhufa*), into a land system in which each peasant household was allocated its own individual plot of land (known as *chefa*).[4] Land still belonged to the state, with the peasant households independently farming the land. Afterwards, according to the amount of land and the yields, taxes were collected and the whole started to resemble a system of private ownership of land. The most representative changes were the Lu Kingdom's *chushuimu*[5] system and the Qin Kingdom's *chuzuhe*[6] system during the Warring States period, when corvée as a form of land rent (i.e., a form of tax) started to be converted into rent paid in kind (i.e., another form of tax). That is, taxes started to be paid in kind by farm commodities such as grain instead of by corvée labor. The reforms of the land system and tax system in these two kingdoms then spread to other kingdoms, one after another. This accelerated the decline of the well-field system and little by little promoted the establishment of a system of private ownership of land (Table 3.2).

3.2.3 Following the Han Dynasty—Privatization of Most Land

At the beginning of the Western Han dynasty, the government adopted the land system of the Qin dynasty. Although land was nominally under state ownership, there was de facto private ownership of land. At the beginning of the Han dynasty the government implemented an enlightened "Huang-Lao,"[7] polity, which [as far as land reforms were concerned] was mainly about reducing corvée and land taxes in order to allow people to recuperate from the traumas of the recent dynastic change

[4]Due to a lack of statistics and contradictory early sources, there has been considerable controversy among ancient and modern scholars about the form, details, and implications of these two systems, especially *chefa*.—trans.

[5]Literally, "start of tax on land per *mu*," this system, promulgated in 594 B.C. in the State of Lu, began the recognition of private land for purposes of taxation. Under the well-field system, communally tilled land had constituted 10% of the land, the produce of which went to the government as a de facto tax; when this system was abolished, tax revenue was changed into a charge of approximately 20% of the yield per *mu* of private land, to be paid in kind.—trans.

[6]Literally, "start of payment in grain," the Qin Kingdom's *chuzuhe* system was roughly equivalent to the Lu Kingdom's *chushuimu* system.—trans.

[7]*Huang-Lao* is a portmanteau word, with Huang referring to Huangdi (the "Yellow Emperor," one of greatest Chinese culture heroes and from highest antiquity, as mentioned in Preface 2 above) and Lao referring to Laozi, of the Spring and Autumn period, and reputed founder of Taoism. *Huang-Lao* was a syncretic philosophy and is seen as a precursor of religious Taoism. Politically, during the Han period the *Huang-Lao* faction at court was in favor of strengthening the power of the feudal kings, in opposition to Confucian and Legalistic factions.—trans.

Table 3.2 Some of the important land reforms of the Spring and Autumn and warring states periods

Kingdom	Year	Content of land reform	Reform instigator
Qi State (1122–221 B.C.)	685 B.C.	Equal-field system, "*xiang di cui zheng*" ("observe the land and decrease the tax demands on it, if need be"); Acknowledged private ownership of land	Guan Zhong (730–645 B.C.)
Lu State (11th century to 221 B.C.)	594 B.C.	*Chushuimu* system ("start of tax on land per *mu*"); Abolishment of well-field system; Acknowledged private ownership of public and private fields	
Qin State (475–220 B.C.)	408 B.C.	*Chuzuhe* system ("start of payment in grain"); Transferring public fields to private hands	
Wei State (407–225 B.C.)	422 B.C.	Priority given to instruction in land fertility; Abolition of previous boundaries of enfeoffment between fields that had been marked by foot paths; Encouragement of unrestrained development of new arable land and industrious cultivation; Development of feudal small peasant economy; Enhancement of production	Li Kui (455–395 B.C.)
Qin State	350 B.C.	Abolition of well-field system and the opening up of footpaths that had marked the boundaries of enfeoffed land; Allowing land transactions, and promoting land privatization	Shang Yang (C. 398–338 B.C.) [Lord Shang]
Qin Dynasty	216 B.C.	Implemented private ownership of land; Landlords and peasant households honestly reported amount of their land and paid taxes accordingly	Qin Shihuangdi (259–210 B.C.) [First Emperor of the Qin dynasty]

Sources Sima Qian, *Shiji*, *Liuguo nianbiao* (*Chronological table of the Six Kingdoms* in the *Records of the Grand Historian*); Ban Gu and Ban Zhao, *Hanshu*, *Shihuo ji* (*History of the* [*Former*] *Han dynasty*, *Treatise on Food and Money*); Zhao Gang and Chen Zhongyi, *Zhongguo tudi zhidu shi* (*History of Chinese land tenure systems*), 2006

and to rebuild the country. Large numbers of landowning peasants worked their own land. Thus, the economy and society recovered quickly.

After proclaiming the *Ernian lüling* (*Laws and decrees of the second year of the dynasty*), which were written on slips of bamboo, the Western (Former) Han dynasty's Emperor Wen (r. 180–157 B.C.) abolished the universal land conferral system. (Previously, all the people in the empire, including commoners, government officials and nobles, had been granted land). This was changed into a system of granting of lands based on military achievement, recognition of social rank, etc.

Although land still nominally belonged to the state, in fact long-lasting private property rights came to include the right of land possession and usufruct. Moreover, peasants had the rights to bequeath and inherit, bestow, and buy and sell their land. The new land system marked an important milestone in the privatization of land during the Han dynasty.

From the perspective of modern property rights laws of the People's Republic of China, throughout the long historic periods of the Spring and Autumn and the Warring States, and from the Qin dynasty to the time of Emperor Wen of the Western Han dynasty, the rights of privately-owned land gradually aligned with the Four Great Powers (i.e., procession, use, benefit, and disposition of property) of state property rights. From this, the system of private ownership of land evolved by degrees and officially became the most important land ownership system during China's feudal period.

According to the research of the scholars Zhao Gang and Chen Zhongyi, although every dynasty had various forms of state-owned lands, the amounts were far less than that of privately owned land.[8]

In a land system that was based on private ownership of land, feudal dynasties intervened a great deal in the land system for the purposes of maintaining stability of agricultural production, expanding the amount of farmland, ensuring land tax revenues, optimizing the relationship between people and the land, restraining land annexation, and maintaining social stability. To summarize my understanding of it based on numerous, obscure historical sources, in the allocation of land and land ownership, the most important interventions were limitations on private ownership of land and the equal-field system.

As for the land tax, three of the major tax system reforms that were closely related to the land tax system over the long course of history were called the "Three Brilliant Pearls." These pearls were: the double-taxation system of the Tang dynasty;[9] the single-whip tax reform[10] of the Ming dynasty; and the Qing dynasty's

---

[8]Zhao Gang and Chen Zhongyi, *Zhongguo tudi zhidu shi* (*History of Chinese land tenure systems*).

[9]Called *liangshui zhi*, the double-taxation system was another early attempt to incorporate the corvée labor tax (i.e., the poll tax) into a land tax, thus simplifying the tax paying process for government administrators and taxpayers alike. Further information is provided on the double-taxation system in Chapter 3, Sect. 2, Subsection 5.—trans.

[10]This curious name for a tax law is a pun on its official name, *yitiao bianfa* (literally, "single [unified tax] reform law"). It was changed in popular parlance to *yitiao bianfa* ("single-whip law"), with the word *biàn*, ("reform") being replaced by the word *biān* ("whip"). This implies that the new, unified tax reform was still a whip used to beat the populace, albeit only a single whip.—trans.

tanding rumu[11] tax system. As for land operation systems, with the exception of the traditional owner-cultivated methods, methods such as tenancy and permanent tenancy developed. Almost all dynasties imposed restrictions on land transaction systems and restrained land annexation, but in general, over time there was a tendency to relax these restrictions and standardize them (Table 3.3).

3.2.4 How Land Has Been Allocated Since the Han Dynasty

The relationship between man and the land and the rational allocation of land were the two most important land system issues that every dynasty had to deal with. The state-led land systems in feudal societies underwent changes from limited land ownership systems to the equal-field system. Ultimately, these all withered away. The system of limited private ownership of land was designed to rationally allocate the workforce and fields, safeguard the state's tax and corvée systems, and suppress annexation of land.

Because there was insufficient land to allocate to all, Emperor Wen of the Western Han dynasty was forced to abolish the allocation system and change it to one in which state-owned land was granted to retired soldiers according to three levels based on their military achievements and social status. Of the nineteen ranks of the nobility, those accorded the high title of Marquis Within the Pass were granted the most land, ninety-five *qing* (one *qing* being equal to 15 *mu*). In order of descending rank, Grand Officers of the First Order was awarded one and a half *qing*, while those with no aristocratic rank, such as rank-and-file soldiers and commoners, were granted one *qing*. Those ministers of justice who were guilty of only minor offenses, as well as court jesters, had fifty *mu* bestowed on them. Prior to Emperor Wu (r. 141–87 B.C.) of the Western Han dynasty, one *mu*, called a "little *mu*," was equal to a strip of land one pace [i.e., a double-step] wide and one hundred paces long. One little *mu* was approximately equivalent to 0.3 of today's *mu*. Emperor Wu of the Western Han unified field measurements and defined the *mu* as an area one pace wide and 240 paces long, called a "big *mu*," which was approximately equal to 0.7 of today's *mu*).

Land conferred to individuals belonged to them and could be passed down to the next generation or could be bought and sold. The state thus acknowledged private land and allowed land transactions.

On the basis of the private ownership land system, the Han dynasty issued an order called the *xiantianling* ["imperial order decreeing limited private ownership of

[11]The *tanding rumu* [literally, "apportion the poll tax as part of the land tax"] system was yet another tax reform that abolished the poll tax by merging it with the land tax. It made taxation easier by allowing both poll and land levies to be paid in money, and not one in money or corvée service and the other in kind. It abolished poll taxes, gave relief to the poorest peasants, and controlled personnel. Moreover, it advanced the development of industry and commerce and was important in reforming the corvée and tax system during the Qing dynasty.—trans.

Table 3.3 Important land systems implementations or reforms during the feudal period

Classification of system	Name of system	Period of implementation or reform	Leading figures	Main content
Land allotment and ownership system	Limited-field system	Western Han, Western Jin	Emperor Wen of the Western (Former) Han dynasty; Sima Yan [known as Emperor Wu, first emperor of Western Jin dynasty]	Western Han: *xiantian ling* ["imperial order decreeing limited private ownership of land"]; Western Jin: *zhantian* ["occupied-field"] system
	Well-field system	Northern Wei to Sui, mid Tang	Emperor Xiaowen of Northern Wei; Emperor Wen of Sui; Emperor Gaozu of Tang; Came to an end with Emperor Dezong of Tang	Peasants granted state-owned land of two types: Public fields [which reverted to the state] Private fields [which could be bequeathed]
Reform of land tax system	Double-tax system	Tang dynasty	Dezong Emperor of Tang; Tang dynasty Chancellor Yang Yan (727–781)	*Zuyongdiaozhi* [triple-tax system] [levying grain, labor, and cloth from peasantry] changed into double-tax system; Poll taxes and corvée taxes merged into land and property tax
Reform of land tax system	Single-whip tax system	Ming dynasty	Jiajing Emperor; Wanli Emperor; Zhang Juzheng (1525–1582) [Grand Secretary] and others	Taxable items such as poll taxes, corvée, etc., merged into Sand tax; Paying taxes in kind gave way to paying taxes using money
	Tanding rumu tax system	Qing dynasty	Kangxi Emperor; Yong-zheng Emperor	Payments in lieu of corvée service included in tax payments; Poll taxes merged into land tax; Taxes levied according to amount of farmland owned

(continued)

Table 3.3 (continued)

Classification of system	Name of system	Period of implementation or reform	Leading figures	Main content
Land trading system	–	Successive dynasties	–	In general, private land was free to be bought and sold; State-owned land had restrictions; Annexation and attempts to inhibit annexation were evident throughout this period
Land management system	Owner cultivated; Land tenancy system	Land tenancy system started in Tang dynasty; Successive dynasties developed and perfected it; Permanent tenancy system peaked during Quing dynasty	–	In addition to owner-cultivated method, tenancy system developed; Pessants who were landless or had little land leased land from landlords and paid rent

land"]. This instituted a system under which the amount of state land conferred on a demobilized soldier or on an individual could not exceed a corresponding quota set for their social status. This was the essence of the *xiantian* ("limited private ownership of land") system. Although it was not implemented thoroughly, in general it ran throughout the Han period.

During the Western Jin dynasty in the first year of the Taikang reign period,[12] the *zhantian* system[13] was actually a system that limited private ownership of land by establishing rules for a limited quota of lands to be allocated to people according to their social positions and circumstances. Male commoners of the Western Jin dynasty were allocated 70 *mu* and females 30 *mu*. For able-bodied males, taxes were levied on 50 *mu*, and for able-bodied females, taxes were levied on 20 *mu*. Officials of the first rank could receive 50 *qing*, and each rank below that was reduced by 5 *qing*, until the 9th rank received only 10 *qing*.

---

[12]This was the third reign period (from A.D. 280–289) of the Western Jin dynasty's Emperor Wu (r. 265–290) [during this period of Chinese history, reign periods were changed frequently, often in accordance with divination or astrological advice]. His actual name was Sima Yan (236–290), the founding emperor of the Western Jin dynasty.—trans.

[13]Literally, "occupied-field system"; this was one of the precursors of the more famous equal-field system.—trans.

After the Western Jin dynasty collapsed and the Eastern Jin dynasty moved its capital to the south, the northern part of China was consumed by the flames of war. Fluttering on top of the city walls, the flags signifying who was king seemed to change constantly. The Sixteen Kingdoms emerged chaotically on the stage of history. The common people wandered destitute and homeless. A great amount of land was abandoned and reverted to undergrowth.

After the Northern Wei dynasty unified North China in 439, the first and foremost problems they faced were overgrown, deserted farmland; agricultural decay; and a paucity of tax revenue. To resolve these problems, the Northern Wei dynasty creatively issued one of the most important land system reforms in history—the equal-field system.

The equal-field system literally meant an equal allocation of land. However, it was not that simple. Under the equal-field system, the country conferred state-owned land on peasants based on the number of family members and labor conditions. There were two kinds of land: private land that did not need to be returned to the government and private land that had to revert to the government.

The Xiaowen Emperor[14] of the Northern Wei dynasty followed the advice of his minister, Li Anshi (443–493), and put into practice the *Law of Equalized Field Allotments* [which became known as the equal-field system]. It divided land into state-owned and privately owned. Based on the usage of land, grain and hemp fields were state-owned, while mulberry fields and fields that were taxed to provide for retired officials were privately owned. According to the law, men older than fifteen years of age could receive 40 *mu* of grain fields, while women could receive 20 *mu*. While these grain fields were state-owned, peasants could cultivate them until they either died or reached seventy years of age, at which time their fields had to revert to the state. Probably very few people during that time reached the age of seventy. Moreover, adult males could also receive 20 *mu* of mulberry land (this was land to plant mulberry or jujube trees on, analogous to today's economically valuable forests), which did not have to revert to the state. Mulberry groves were private lands that the state allocated to peasants.

In addition to the fact that slaves and servants could get land for farming, what is astonishing to modern people is that a full-grown ox could also be granted 30 *mu* of land. However, given the social background of that time, it is understandable; oxen were a main means of production in ancient times. The logic was that more land could be plowed with them, so of course oxen should be given some land. The Northern Qi and the Northern Zhou dynasties carried on the practice of the equal-field system.

[14]This was Tuoba Hong (467–499), who in 496 founded the Northern Wei dynasty and became the Xiaowen Emperor. The Tuoba were a non-Han ethnic group of northern, herding nomads who were Sinicized and came to rule part of Northern China.—trans.

The Sui dynasty broadly promoted the equal-field system, with peasants receiving more land per capita. Each adult male was awarded 100 *mu*, 20 *mu* of which was permanently held and could be passed on to descendants. The remaining 80 *mu* were designated *koufentian*.[15] Women received 40 *mu* of *koufentian* land (during the Sui and Tang dynasties, one *mu* was approximately equal to 0.8 of today's *mu*). After the death of the recipient, the *koufentian* land reverted to the state. Perpetually held land was de facto privately owned land, while *koufentian* was state-owned. Slaves who were allowed to participate in the system received the same amount of *kuofentian* land as commoners, but the number of slaves who could receive land was limited and was based on the slaveholder's rank of nobility or governmental grade (if the slaveholder was an official). Fully grown oxen were allotted 60 *mu* each, with families being limited to four head of oxen receiving land.[16]

3.2.5 "Three Bright Pearls" in the History of Taxation

The Tang dynasty inherited from the Northern Wei dynasty its equal-field system. In addition, the Tang dynasty promulgated the *zuyongdiao* ["triple-tax"] system used for levying grain, labor, and cloth (or other local items) from the peasantry. This stipulated that able-bodied men who received farmland from the government must pay a land tax in grain; this was called *zu*. Furthermore, they must perform corvée; this was called *yong*. In lieu of performing corvée, one could choose to pay money to the local government with which it could then hire a substitute; this payment was also known as *yong*. Finally, they must pay a special household tax (paid in locally produced specialty products, such as silk cloth); this tax was called *diao*. The *zuyongdiao* "triple-tax" system played an important role in developing agricultural production and thus promoting the prosperity of the Tang dynasty.

During the mid Tang period, as a result of population growth and land annexation, the state had insufficient land to confer on the peasants, so the equal-field system could no longer be implemented and the *zuyongdiao* triple-tax system lost its foundation. During the reign of the Dezong Emperor of Tang Dynasty (742–805; r. 779–805), the Prime Minister Yang Yan (727–781) created the double-taxation system, abolishing the equal-field system: "No longer will the government differentiate between original residents of a locale and non-natives who have migrated there. Provided that they have property and land, they all will be counted as locals

[15]I.e., lands that were parceled out based on the number and kind of family members. The formula allotting farmland to households members included allotments for the elderly, the sick, the disabled, widows, etc.—trans.

[16]Wang Zhongluo, *Sui Tang Wu Dai shi* (*shang*) (*History of the Sui, Tang, and Five dynasties*, vol. 1).

and put on the tax rolls. They will all pay a land tax that incorporates in it the corvée [i.e., labor or poll] tax .... Taxes are to be paid in the autumn and summer."[17] Because of this, it was called the double-taxation system.[18]

The transformation of the Chinese tax system from the *zuyongdiao* triple-tax system to the double-taxation system was an important reform. It combined various taxes and caused the ancient Chinese poll tax to be changed into a land and property tax. At the same time, it abolished the equal-field system that had been the method that the state had controlled land distribution.

The Ming dynasty formulated rules and laws to regulate the land system through the *Da Ming Lü* (*The Great Ming code*).[19] The following three laws were the most important of these.

First, it established a strict system for the registration of land property rights and tax payments. This stipulated that, "Every household must fill out a document based on its actual situation. It is strictly forbidden for officials and gentry to collude, take over control of land, or tamper with the process in anyway." Furthermore, it stated that, "Without exception, all the following officials and persons in *lijia*[20] neighborhood organizations who are guilty of the following violations will be put to death: those who collude; those who withhold truth or commit fraud; those who falsely report that private land is included as part of public land; those who sneakily delay transfer of ownership of land in order to avoid taxes; those who are involved in any form of forgery of documents; moreover, those who shortchange the amount of grain tax that they owe. Households that conceal the facts shall have their patriarchs executed. The whole of such families shall be transported beyond the pale of civilization."[21]

Second, in order to prevent the evading of the land tax by subterfuge, laws emphasized that in land transactions all land taxes had to be paid and corvée performed before transfer of ownership was allowed.

Third, for the first time in history in the form of law, the quality and type of the government-owned land underwent clear-cut classification. Moreover, in official land registers, the land belonging to officials was differentiated from that belonging to individuals. During the Ming period it was forbidden to change the category of

---

[17]Wang Pu (922–982), editor, *Tang huiyao* 唐会要 (*Essential documents and regulations of the Tang dynasty*). This was compiled in 961 and is the earliest extant book of the *huiyao* (*Essential documents and regulations*) genre.—trans.

[18]There are at least three possible derivations of the name for double-taxation. One, mentioned here, is that taxes were collected twice a year. Another explanation is that the taxes were composed of a household tax (*hushui*) and a land tax (*dishui*). A third possible derivation is because it consisted of a combined tax on landed property and on profits or other property.—trans.

[19]This was short for *Da Ming lü jijie chulie* (*Legal code of the Great Ming, with collected explanations and appended substitutes*), enacted by the founding emperor of the Ming dynasty, Zhu Yuanzhang (r. 1368–1398) in 1373.—trans.

[20]The Ming dynasty's *lijia* system organized neighborhoods into 110 households constituting one *li* or "neighborhood"; this was further subdivided into ten groups called *jia*.—trans.

[21]Sun Zhilu, (fl. Qing dynasty), *Er shen yelu* (*Unofficial history of* 1368–1644 [*i.e., the Ming dynasty*]).

state-owned fields into that of non-state-owned fields [i.e., to be sold to private individuals], other than exceptions permitted by government laws. It was also forbidden to change the classification of any state-owned fields.

The taxes on the use of state-owned fields were different from those paid on privately owned fields; taxes on different categories of state-owned land differed as well. In general, it was higher than that on private land. This guaranteed that the government maintained absolute control of a sufficient proportion of the land and obtained lasting, high taxes from state-owned fields to meet the needs of government finance and to inhibit land annexation.

During the reigns of the Ming dynasty's Jiajing (r. 1522–1566) and Wanli (r. 1573–1619) emperors, the rulers promoted the single-whip system to simplify the tax system. The general idea was to amalgamate the poll tax and corvée labor tax, etc., into a single tax, the land tax. The second aspect was that the method of paying taxes was shifted from payment in kind to payment primarily with money. Because of these two changes, the single-whip tax system has been regarded as the most important tax reform in Chinese history.

At the end of the reign of the Qing dynasty's Kangxi Emperor (r. 1662–1722) and during the reign of his successor, the Yongzheng Emperor (r. 1723–1735), the tax reform known as the *tanding rumu* tax system was implemented. This was done in order to alleviate class conflict caused by land annexation, to lighten the tax load on the populace, and to simplify the form of land taxes. In the fifty-first year of the reign of Emperor Kangxi [i.e., in 1712], the Qing government promulgated the slogan "Adding new family members will not increase taxes." With the number of people subjected to the poll tax stabilized, it laid the groundwork for implementation of the *tanding rumu* system. The Yongzheng Emperor formally implemented this reform, merging the tax paid in lieu of corvée [i.e., a poll tax] with the land tax. The poll tax was incorporated into the land tax and was then collected as part of the tax on the number of *mu* under tillage, thereby completely merging the corvée and the land tax systems.

The attentive reader may ask, "Didn't the Tang dynasty's double-taxation system and the single-whip system of the Ming dynasty already take the different kinds of taxes, especially the poll tax and corvée labor, and incorporate them into the land tax?" Yes, that is correct. This conundrum can be explained by what Huang Zongxi (1610–1695), an intellectual of the late Ming and early Qing dynasties, called "the harm caused by accretions of [tax] reforms." The main idea here is that in attempts to lighten the tax burden, previous historical tax system reforms would merge and simplify taxes, but then add new kinds of taxes; thus, the total tax burden would actually become heavier instead of lighter. Based on the insights of Huang Zongxi, the modern scholar Qin Hui (1953–) summarized them as "Huang Zongxi's Law."[22] Nowadays, because of modern reforms, farming is finally free of the

[22]Qin Hui, *Nongmin jianfu yao fangzhi Huang Zhongxi dinglü xianjing* ("*When lightening the burden on farmers, beware of the trap of Huang Zongxi's Law*"), *Zhongguo jingji shibao* (*Chinese Economic Times*).—trans.

problem of those ancient taxes. But what about all the other taxes that had been added such as land taxes, industry and service taxes, and income taxes?

3.2.6 Buying, Selling, and Leasing of Land Became the Norm

Since the beginning of the Qin and Han dynasties, the buying and selling of land has marked the private ownership land system. In the flourishing Tang dynasty, private contracts for transacting land were recognized by government. The buying and selling of land changed from a system restricted by government into an unrestricted system, leading to a thriving market in real estate. With the explosion of private ownership of land and landlord estates, a new kind of land management method—leasing land to tenants—started to develop, resulting in the gradual loosening of the fetters that tied laborers to the land.

Due to insufficient land, the Song dynasty adapted a land policy of "No establishment of a land system" (i.e., no continuation of the equal-field system of the previous Tang dynasty) and "No suppression of land annexation" (as recorded in the *Songshi*, *Shihuozhi* [*History of the Song dynasty*, *treatise on food and money*]). Instead, the Song dynasty acknowledged and protected the legitimacy of property rights for private land and the commercial nature of land. It allowed land to be transacted in accordance with economic laws. The state would no longer intervene in land matters and no longer inhibit land annexation.

In the Song dynasty, with the exception of state-owned land, the transactions of private lands only required general procedures. First, land contracts had to be submitted to the local government and receive approval by getting an official chop (this sort of contract, which was called a vermillion contract because of the color of ink used for the chop, was in effect a registered deed). Then land tax had to be paid. After that, the land had to be registered in the government deed books. Finally, the buyer assumed responsibility for payment of land taxes for the piece of property. Only then was the transfer in land ownership considered complete.

The land system during the Song period supported the development of private ownership. However, it also tolerated the annexing of land by local despots, accelerating the concentration of landholdings and causing an expansion in the size of country estates. At that time, this relatively open and free land-management method—the tenancy system—came close to perfection. It became second in importance to the management of land-holding peasants. The reason was that land annexation caused a large number of peasants to lose their land. The system of separating people with land and property from those without was used in the Northern Song dynasty and helped the tenancy system to be established legally.

The land system of the Yuan dynasty did not progress much. Two different land ownership and taxpaying systems formed, one in the north and one in the south. In the north, the land was mainly stated-owned or was undeveloped land use to station army garrisons that were expected to clear the land and grow their own food. State-owned land was leased out to peasants who became tenants. In the south, there was a system of private ownership of land.

During the Qing dynasty, state-owned farmland, called *guantian*, was divided into banner land[23]; land for military garrisons to cultivate; communal land used to support local schools; and grazing land. Farmland not owned by the government was called "peoples' farmland" [i.e., private land]. Landlords owned the vast majority of "peoples' farmland," who then leased it to tenants. Those peasants who had large holdings, and those who owned some land but had to rent more to have enough, merely possessed a small portion of the total land. After the Qing dynasty's Qianlong Emperor, the system that forbade state-owned farmland to be transacted was abolished. As a result, the amount of state-owned land continued to decrease and private land greatly increased.

Accompanying the unprecedented development of buying and selling of land, the landlord economy prospered. People's relationship with land also experienced great changes. During the Qianlong Emperor's reign, permanent tenancy in Zhejiang, Jiangsu, Hubei, and Hebei provinces developed and entered a flourishing period.

There appears to have been three impetuses leading to permanent tenancy: first, the opening up and reclaiming of wastelands; second, development of the rent deposit system, in which rent was paid in advance by tenants to be returned by landlords at the end of the lease period; and finally, the development of land mortgaging systems.

The right of permanent tenancy was a kind of independent property right. Peasant households who enjoyed this right were at liberty to decide what crops to plant. If they did not wish to cultivate the land themselves, they could transfer their rights to others, including subleasing to other tenant farmers, bequeathing their rights to their children, or pawning or selling their rights on the open market. When the right of permanent tenancy was sold off, the price was decided by agreement between buyer and seller.[24]

Permanent tenancy separated landlord ownership rights (i.e., the right to collect rent from tenants) from those of the tenant (i.e., tenants' rights to use the land and pay rent in the form of grain). At the same time, these rights of tenant farmers were marketized, and there emerged forms of mortgage, permanent sale, and pawning of land. This allowed peasants who had lost land to be able, in multiple ways, to promptly be reunited with it. Hence, the land utilization rate greatly increased, promoting the development of social production and leading to a dramatic expansion of land markets.

---

[23]This was land reserved for members of the Manchurian ruling class, who belonged to one of eight "banners" [i.e., ruling clans].—trans.

[24]Zhao Gang and Chen Zhongyi, *Zhongguo tudi zhidu shi* (*History of Chinese land tenure systems*).

In the meantime, the fixed-rent system began to develop, which was premised on farmers having high and stable yields. Under this system, in general the landowner did not meddle with farming, so the tenant was free to make his own decisions. This prevented the landowner from interfering in production. The basis of the fixed-rent system was that the tenant obtained more freedom to operate. In this way, all managerial rights and some ownership rights were shifted from the landowner to the tenant farmer. This was the most important progress made in the feudal tenancy system and was its central characteristic during the Ming and Qing periods.

3.3 The Evolution of the Rural Land System in the People's Republic of China

Since the founding of the People's Republic of China, the Chinese land system has experienced three great changes. The first was agrarian reforms at the beginning of the 1950s; under the slogan "land to the tiller," this ended the feudal ownership system. Farmers thereby acquired private ownership of rural land.

The second change came in the mid 1950s, along with the campaigns of agrarian collectivization and the People's Communes. The land property rights system evolved from one of farmers' private ownership of land into one of collective ownership. The farmland then belonged to farmers as a whole rather than to individual farmers. The land management system evolved at first from farm household operations into cooperative management and then later into unified, collective management.

The third and final major change came after the rural reforms in the early 1980s. The household contracted management system was implemented in which land rights still belonged to the collective, while farmers were able to enjoy management of farmland by contracting it from the collective, of which they were a part. This was the separation of two rights.[25] Farmland was changed from collective management to farm household management (Urban land continues to be state-owned, but this will not be discussed in this book.)

At the current time, there are changes in progress that involve the "Separation of the Three Rights." That is, the separation of the rights of: one, rural land ownership; two, contracted rights; and three, operational rights. These taken together may be seen as a fourth big change, but at present, the results of this change are still unclear.

[25]This is in essence the principle of separating the right of enterprise ownership from that of management.—trans.

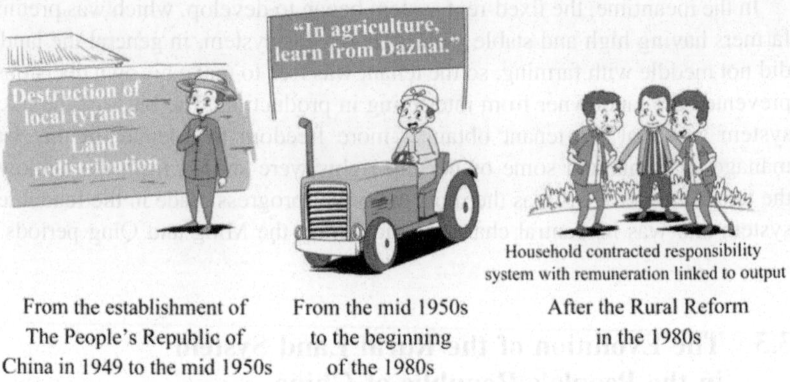

Evolution of the People's Republic of China's Land System

[Dazhai in Xiyang County, Shanxi Province, from 1964 to the late 1970s was popularized as a model agricultural brigade in China.]

3.3.1 Farmer-Owned Land, Farmers Managed Cultivation (from Establishment of the PRC to the Mid 1950s)

After the establishment of the People's Republic of China in 1949, the Chinese Communist Party began nationwide agrarian reforms in order to abolish the feudal landlord system. On June 28, 1950, the Central People's Government of the PRC, in accordance with the new situation resulting from the liberation of the country, promulgated the *Agrarian Reform Law of The People's Republic of China*, which included three important items:

First, there was the abolishment of the exploitative feudal land ownership system of the landlord class, and the implementation of a system of land ownership by farmers. Second, all land that was confiscated and collected, together with other production materials (with the exception of those that were given over to state ownership under this law), was to be taken over by county farmer associations. Then it was to be distributed fairly and reasonably to poor farmers who were landless or who had only a little land and who lacked production materials. Finally, after the completion of agrarian reforms, certificates of land ownership were to be issued by the people's government, acknowledging the rights of freedom of management, transactions, and leasing for all landowners.

In the spring of 1953, with the exceptions of some regions of ethnic minorities, agrarian reform was brought to a smooth end. From then on, the notion of "land to the tiller," which the Chinese people had been dreaming of throughout history, at last became a reality. The agrarian reforms allowed 70% of farmers of the rural population who lacked land to receive land at no charge. During this period, farmers enjoyed land and housesite ownership rights and held certificates of property rights.

This new land system produced good effects by stimulating farmers' zeal to develop production and construct new homes.

The Constitution of the People's Republic of China, issued in 1954, spelled out farmers' land rights with even more authority. The nation, in accordance with law, protected farmers' land ownership rights; the nation protected citizens' ownership rights to lawfully earned income, savings, houses, and means of livelihood; and the nation, in accordance with law, protected citizens' rights to inherit private property.

Through the agrarian reform law and the constitution, farmers' land rights were made clear. At this stage, farmers were both land-owners and land managers; the rights of land ownership, management, usufruct, and disposal of property were unified to a high degree. The transfer of land property rights was smooth, without any institutional obstacles; land could be bought and sold, rented and leased, pawned, bequeathed, and inherited—all was permitted. It also included full property rights for housesites and buildings.

3.3.2 *Collective Ownership of Land and Collective Unified Management (from the Mid 1950s to the Beginning of the 1980s)*

In the mid 1950s, in accordance with its socialized economic system, China started campaigns to establish agricultural cooperatives and collectivization. On February 25, 1953, the Central Committee of the CPC issued a "Resolution on the Development of Mutual Aid and Cooperation in Agricultural Production." At first, mutual aid teams were established.[26] Land, other means of production, and products all belonged to team members. Farmers, based on "voluntary participation and mutual benefit," took part in small-scale cooperative production.

On January 8, 1954, the Central Committee of the CPC issued a *Resolution on the Development of Agricultural Production Cooperatives* that indicated a need for "gradual realization of socialist transformation of agriculture, the handicraft industry, and capitalist industry and commerce." This stipulated that: the land of farmers would merge with a cooperative, farmers would be given shares in it, and the cooperatives would manage the land in a unified way. Properties such as farm animals and large farm implements shall be used in a unified, cooperative manner. The land of elementary cooperatives[27] would still be privately owned; yet, farmers would retain the rights to income distribution, if they joined a cooperative.

[26]Mutual aid teams were an elementary form of organization in China's agricultural cooperatives in the early 1950s. These were based on the principle of voluntary mutual aid. A team was usually composed of up to a dozen or so farm households that helped each other out.—trans.

[27]Elementary cooperatives were established in the early 1950s. Distribution of income was according not only to the amount of work each member did but also on the amount of land each member contributed.—trans.

On November 9, 1955, the Standing Committee of the National People's Congress (NPC), in a *Draft of Standard Charter of Agricultural Production Cooperatives*, put forth: (1) It is important to gradually replace the private ownership of the means of production with a collective ownership system. (2) There are two stages in the development of cooperative agricultural production, the elementary and the advanced. (3) In the elementary stage, cooperatives are to be semi-socialistic, as land ownership still resides in the farmers. (4) In management, in the elementary stage, the individual household operation is shifted to a cooperative management system.

On June, 30, 1956, the Third Session of the first National People's Congress of the PRC passed the *Standard Charter of Advanced Agricultural Production Cooperatives*, which stipulated that: agricultural production cooperatives, in accordance with the principals of socialism, should transfer the main production materials from private ownership to collective ownership. Farmers joining cooperatives must transfer means of production such as land, livestock, and large farming implements to cooperatives.

Following extensive promotion of collectivization, by the end of 1956, there were 756,000 agricultural production cooperatives nationwide. Ninety-six point three percent of farm households joined cooperatives; of these, 87.8% joined advanced cooperatives. The socialist transformation of the means of production in rural China was essentially completed.

In 1957, almost all the land that was previously owned by farmers was handed over by force to advanced cooperatives. In order to expand cooperatives and reach the goal of communism as soon as possible, our country promoted agrarian reform by transferring all land to public ownership.

In March 1958, the Central Committee of the CPC passed *Suggestions for Small-Scale Cooperatives to Appropriately Merge with Large Cooperatives*, which urged that, "In those areas with proper conditions, small cooperatives should merge with large cooperatives in a moderate, planned way." This was a harbinger of the People's Communes.

On August 29, 1958, the Political Bureau of the CPC Central Committee, in the enlarged session held in Beidaihe,[28] issued a *Resolution on the Matter of Establishing People's Communes in Rural Areas*, which rapidly popularized People's Communes throughout the country. As a result, by the end of the year, almost all 120 million families had joined over 26,000 People's Communes. The system of People's Communes terminated the system of farmers' ownership of land and formed a land system based on "collective ownership" and "collective management."

[28]Beidaihe is a popular seaside summer resort town in Hebei Province.—trans.

Initially, the general features of the People's Commune system were incorporated in two popular slogans: "First, big; second, publicly owned" and "First, [indiscriminate] equalization; second, [indiscriminate] allocation of resources."[29]

"First big" meant to first make the scale of the People's Communes big; basically, this led to every township having a commune, with thousands of farming households in one commune, and with a high degree of public ownership of land, livestock, and other means of production, including consumer goods.

The slogan "first, big; second, publicly owned" led to the second slogan "First, [indiscriminate] equalization; second, [indiscriminate] allocation of resources." After cooperatives were amalgamated into People's Communes, property was transferred to a higher agency where there was centralized accounting and unified allocation in order to implement egalitarianism. The system of People's Communes, set up within such a short time, was difficult for the more than one hundred million farming households to cope with. The egalitarian distribution system, which took large quantities of communally produced goods and allocated them freely, did extensive harm to manufacturing and to the interests of farming families; this had serious negative effects on farmers' zeal for production.

During the period from 1959 to 1962, the CPC Central Committee undertook several adjustments of the administrative system of the production management system of the People's Communes.

On September 27, 1962, the Tenth Plenary Session of the 8th Central Committee of the CPC officially passed *A Draft Amendment to the People's Communes' Work Regulations*. This took the shape of a rural collective ownership, operation, and management system, a three-tier ownership system with production teams as the basis. That is to say, production teams were the basic units of production and accounting. They were also the owners of land and other means of production. The People's Commune system continued until the rural reforms of the early 1980s.

Some people might mistakenly think that the ideas encapsulated by the slogans "First, big; second, publicly owned" and "First, [indiscriminate] equalization; second, [indiscriminate] allocation of resources" were the total salient features throughout the period of the People's Communes. In fact, after the establishment of the system of "three-tier ownership with production teams as the basis," these two initial slogans no longer served as the main features of the People's Communes. This was because on communes, the basic unit of production, accounting, and means of production were production teams made up of scores of farm households. The egalitarianism in allocation had already changed significantly. The [indiscriminate] equalization and free allocation of resources had also been forbidden by documents from the central authorities that had furthermore demanded that compensation be paid. Meanwhile, the People's Commune system, in which there was an integration of government administration and commune management, had

[29]Mao Zedong called for communes to be bigger in size than the cooperatives, to be publicly owned, and to then take the goods produced in common and distribute them freely, all in order to facilitate the movement toward Communism.—trans.

already loosened somewhat. From production organizations and accounting units, going all the way down to production teams, the main function of People's Communes had changed; they had become more administrative in function and manifested themselves more as administrative units.

Although during the period of the People's Communes the collectively managed main body of production organizations and accounting units were moved downward from the People's Communes toward the production teams, the basic characteristics of the rural land system (which was "collective ownership and collective management") never changed. These basic characteristics continued uninterrupted until the rural reforms of the early 1980s.

3.3.3 Collective Ownership of Rural Land and Land Operation by Farmers (The Period After the Rural Reforms)

After the reforms and opening up policy, comprehensive economic reforms were carried out. In rural areas, a land system was established in which land was collectively owned and farmers made use of it, with the main body of ownership still residing with the farmers' collective. The biggest reform was the land management system that changed from collective management to household contracted land management.

The reforms that brought about the household contracted land management system were not in fact formulated overnight. There had been heated debate in society and among high authorities. In May, 1980, Deng Xiaoping (1904–1997) issued *Talks on Government Policy Regarding Rural Areas*, in which he pointed out: "Since the loosening of government rural policy, there have been some experiments with fixed-output quotas in areas with suitable conditions. These have shown good results and quick changes." This statement, based on reality, unified people's understanding, pushing forward the rural land reforms that had just started.

In September, 1980, the Central Committee of the CPC issued *No. 75 Central Document*, *Several Questions on Further Strengthening and Improving the Agricultural Production Responsibility System*. It emphasized that the collective economy was the basis for our country's rural economy. It also fully acknowledged the responsibility system of specialized contracts and the linking of payment to output. The document affirmed the practice of "households' fixed-output quotas," regarding it as a necessary measure to develop production and solve the problems of inadequate food and clothing.

In January, 1982, the CPC Central Committee forwarded the *Summary of the National Rural Work Conference* (the CPC Central Committee's *No. 1 Central Document* of that year) that for the first time clearly confirmed the socialist nature of household-based fixed output quotas. It pointed out that, "The current various responsibility systems include labor-quota remuneration for certain kinds of short-term contract work, specialty-work contracts with payment linked to output, production-related labor contracts, "guaranteed-production households" and "guaranteed-production teams,"[30] and the "guaranteed-work households" and "guaranteed-work teams."[31] All of these are production responsibility systems of the socialist collective economy." This document dispelled worries of people, and promoted the great development of the "guarantee-production households" system, with its household-based fixed-output quotas.

On December 4, 1982, the Fifth National People's Congress adopted the Constitution of the People's Republic of China.[32] For the first time in constitutional form, it was indicated clearly that urban land belongs to the state and farmers collectively own rural land. It stipulates (in an excerpt from Article 10) that, "… with the exception of the land that belongs to the state by law, all rural lands and suburban lands belong to collectives. Housesites, private farm plots, and private hills or mountains also belong to collectives."

On December 31, 1982, the Political Bureau of the Central Party Committee passed *A Number of Issues Regarding Current Rural Economic Policies* (the Central Committee's *No. 1 Central Document* of 1983). This document further affirmed the household contracted responsibility system [with remuneration linked to output]. It indicated that the stabilizing and improving of the agricultural production responsibility system was a main task in current rural work. It considered that management methods that combined decentralized with unified management possessed extensive applicability. Therefore, it requested comprehensive promotion of the household contracted responsibility system. Consequently, household contracted management entered a stage of stable development. On the basis of this stability and development, rural economic reforms have continued to intensify.

On June 25, 1986, the 16th Session of the Sixth Standing Committee of the NPC adopted the *Land Management Law of The People's Republic of China*, which explicitly indicated that the household contracted responsibility system in rural areas was in essence the collective ownership of the economy by the socialist working class. It established the legal status of the household contracted responsibility system in the form of law and further enriched the content of land contract policy at this stage. With this, our country's rural land contracted management system was finally established and the rural land system entered a new era.

The household contracted management system enormously mobilized China's huge numbers of farmers and rapidly displayed its tremendous superiority over collective agriculture. This new system, along with the widespread use of chemical fertilizer, have caused extraordinary growth in farm production; food production [i.e., *liangshi*] volume skyrocketed from 305 million tons in 1978 to 407 million tons in 1984.

[30]These two types of work organizations agreed to produce a fixed-output quota on land provided, keeping any surplus and paying fines on any shortages.—trans.

[31]These two types of systems did not have a fixed-output quota, but agree to payment of taxes and sharing of output with the collective.—trans.

[32]This was actually an adoption of the fourth and current version of the Chinese Constitution (the others versions were adopted in 1954, 1975, and 1978); this newest version has been subsequently amended in 1988, 1993, 1999, and 2004. —trans.

3.4 How Can Farmers' Collective Land Ownership Be Perfected?

If you were a farmer, what rights would you have to your land? To make it more complicated, suppose you were a farmer who had left your rural village to do manual labor or to do business. Would your land rights to the farm that you left behind change? To complicate it even more, suppose that you were now a teacher or civil servant; your occupation and identity have changed, but you still have your share of land by the contract land agreement for the years stipulated in the contract. What then would your land ownership rights be? (Table 3.4).

Table 3.4 Land rights

Holder of rights	Name of rights	Category of rights	Subdivisions of rights	Flaws in rights	Direction of reforms
Fanners' collectives	Land ownership rights	Rights of ownership	Rights of possession, usage, usufruct, and disposal	Property rights not perfected; Unstable rights of ownership; Inadequate disposal rights; Unclear membership rights; Obscure legal principles; Deficient rules governing division of income	From incomplete to complete; Restrict expropriation of land; Reform the system of requisitioning of land; Open up direct markets for collective land; Define members' rights; Verify collective status and boundaries; Establish a complete distribution mechanism
Farmers	Contracted land management rights	Rights of Usufruct	Rights of possession and usage	Unstable; Frequent adjustments; Interference with production; Coerced land transactions	From unstable to stable: Clarify land rights; Adopt long-term policy;

(continued)

Table 3.4 (continued)

Holder of rights	Name of rights	Category of rights	Subdivisions of rights	Flaws in rights	Direction of reforms
				and expropriation of land; Short contract periods	Extend contracted land management period; Standardize land transactions and procedures for expropriation of land, etc.
Farmers	Contracted land management rights	Rights of Usufruct	Usufructuary rights, including; Production; Land transactions; Compensation for expropriated land; Drawing dividends if an equity investor	Low compensation for land expropriation; Difficulty in capturing gains on appreciation of land value	From insufficiency to sufficiency; Refer to land market value to determine compensation; Become equity investors and share profits or dividends
			Right of disposition, including; Transactions; Mortgaging; Bequeathal and inheritance; Withdrawing from land ownership rights and obligations, etc.	Transactions unstandardized; Limited rights to mortgage; Obscure rights of bequeathal and inheritance; No method to relinquish land ownership rights and obligations	Standardize transactions; Lift the ban on mortgages; Permit inheritance; Permit compensation for relinquishing land ownership rights and obligations

Note: The right of farmers to use land for housesites is also one of the important land-use rights; this is not discussed in this work. Readers interested in this subject are referred to the work that I edited, *Wanshan yu gaige nongcun zhaijidi zhidu yanjiu (Research into the reform and refinement of the rural homesite system)*, China Agricultural Press, 2011

It all comes down to this question: what are the details of rural land rights and who can enjoy those rights? This is a basic yet vital question because it involves individual interests that directly affect 880 million farmers, whether one is a farmer who stays on the farm or migrates to the city to do manual work. The answer is, there are two main bodies of rural land rights: those of collectives of farmers and those of individual farmers. There are two major kinds of rights: property rights and usufructuary rights. These two rights can explain the basic framework of rural land rights.

Before the issuance of the *Real Right Law* in 2007, legal definitions of land rights were comparatively vague. For example, the legal relationship between collective land ownership rights and contracted management rights lacked legal definitions, as did the legal content of contracted land management rights, etc. The *Real Right Law* defines land rights using legal statutes. Land rights are a property right of real estate.[33] Farmers possess collective ownership of rural land and they have the right to use the land. This mainly includes two usufructuary rights—contracted land management rights and housesite rights.

Among land rights, the right of ownership is at the core. This is the basis from which other rights emerge and development. The special characteristic of China's land system lies in that land ownership is divided into two categories: state-owned and collectively owned lands. In rural areas, land mainly belongs to collectives, while state-owned farmland belongs to the government. As mentioned above, the Constitution says, "… with the exception of the land that belongs to the state by law, all rural lands and suburban lands belong to collectives. Housesites, private farm plots, and private hills or mountains also belong to collectives."

According to Article 58 of the *Real Right Law*, collectively owned movable and immovable property includes: lands, forests, mountain ranges, grasslands, wastelands, and tidal lands that belong to the collective by law. Farmers collectively have land ownership rights, signifying that they collectively enjoy the rights of possession, use, benefit, and disposal of such lands according to law. Nevertheless, due to system design, farmers' collective land ownership is incomplete. As a result, in reality a number of problems still exist.

[33]The *Real Right Law* stipulates that "real right" refers to the exclusive right of direct control enjoyed by the holder according to law over a specific property (both movable and unmovable property), including ownership, usufructuary rights, and guaranteed property rights for security. The owner of real or movable property has the right to possess, use, seek profits from, and dispose of real property or movable property according to law. The owner of real property or movable property has the right to establish a usufructuary right or real right for security over the real property or movable property. The holder of usufructuary rights has the rights to possess, use, and seek profits from the real property or movable property owned by others, according to law.

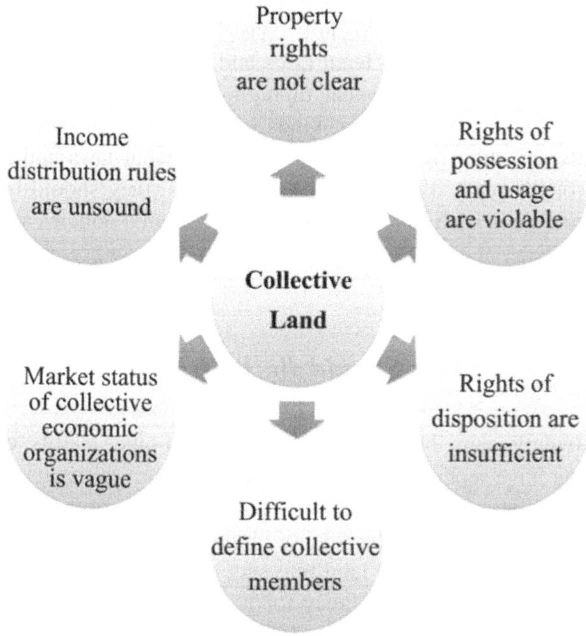

There exist many kinds of problems with collective land

3.4.1 Rural Land Should Belong to Farmers Collectively

Rural land is in the possession of farmers' collectives and the individual farmer is a member of his collective. A great many farmers and grassroots cadres are by no means clear about to whom rural lands actually belongs. My father once held the position of village accountant. He considered that rural land belonged to the country. In 2011, I was in charge of research on the transactions of farmland when we surveyed 669 farm households in twenty-five provinces. Unexpectedly, 44.7% of farm households considered that contracted land belongs to the state while farmers only use it. Another 14.7% of them had no idea about this. According to a survey conducted by the scholar Shi Qinghua, 46.6% of farmers considered that farmland belongs to the state. Not only are farmers like this, but also even some basic-level cadres are muddled-headed when it comes to this question. Therefore, it is very easy for people to mistakenly regard all farmers' collective lands as state-owned.

Because both government officials and farmers are unclear about the ownership of rural land, in some places it is easy to requisition and occupy rural land and abolish villages or consolidate them with other villages on a large scale. Due to the lack of sufficient legal knowledge, plus the fact that change of ownership is a common phenomenon, the awareness that farmers' collectives own the land is getting weaker. Farmers thus do not consider that the land is their own. Nevertheless, after understanding legal rules and regulations, they frequently have epiphanies that cause them to appeal to higher authorities for help. Ergo, it is necessary to use extensively different kinds of media to let farmers clearly know their land rights and their own interests, as well as what laws and regulations to abide by. As for the party officials of different levels, they should be educated to understand regulations, such as land ownership and related policies and the exercise of public rights, etc.

3.4.2 *Land Requisition Should Be Restricted*

Over the past twenty years, because of requisition and other factors, collective land ownership rights have repeatedly been violated. Collectives' rights to possess and use land are unstable and their rights of usufruct have been damaged. If it is necessary to requisition land for the sake of public interest, then farmers will regard changing their collectively owned land into state-owned land as their responsibility and duty, if they are compensated fairly for it. However, if in the name of public welfare, farmers' land is requisitioned to be used in large commercial construction, with low compensation paid to farmers and with great profits accruing to developers, this is a violation of collective land ownership rights.

For the past few years, the country has been attempting to refine the land requisition system for construction purposes. To do that, it has put forward policies such as strict demarcations between land used for the public good and that used for commercial construction, as well as narrowing the scope of land requisitions and improving compensation. However, the large surge in land used for commercial construction has not abated at all. Even though systems to protect farmland are extremely strict and requisition is increasingly difficult, collective land ownership rights have suffered new assaults.

[In order to get around increasing restrictions on acquiring rural land for construction,] quite a number of places have started changing the ways that they go about getting such land. Now, in the name of constructing a new countryside, they concentrate construction of dwellings in one small area, exchange housesites [e.g., offering farmers apartments in a centralized area in order to take over their housesites], or abolish boundaries between villages to dissolve some of them or merge them with other villages. The resulting surplus of land is then used for commercial construction. The consequence is that collectives cannot guarantee their rights of possession, use, or usufruct of the land.

Henceforth, land ownership rights of collectives should be safeguarded by reasonable constraints on government land requisition and on any village renewal that involves the demolition, merger, or consolidation of villages. There should also be reform of the existing legal system of government land requisition by clearly defining "public benefit," narrowing the scope of land requisition, and the marketization of rural land used for construction.

How can the scope of land requisitions be reduced? How can land for public benefit and for commercial use be defined? Moreover, how can requisitioning land be prevented that claims to be for public benefit but is in fact for commercial purposes?

At the Third Plenary Session of the 18th Central Committee of the CPC in 2013, the proposal was made to "expand the scope of compensation for use of state-owned land, and decrease the allocation of land that is not used for public benefit." The 2001 issue of the *Land Allocation Catalog* by the Ministry of Land and Resources should be used as the basis to strictly draw up a catalog of land used for public benefit. As for energy industries (e.g., oil, coal, natural gas, power transmission, and hydroelectricity) and transportation industries (e.g., highways, railways, aviation, and waterways), it is neither easy nor necessary to distinguish between those that require state key support and those that do not. Furthermore, enterprises in energy and transportation industries process a certain monopolistic nature, and the main thrust of their marketing and management is to make profits. Therefore, costs of land requisition ought to be internalized as part of their business costs. It is not right to harm the interests of vulnerable farmers in order to subsidize powerful energy and transportation enterprises.

There should also be a catalog, with strict definitions of land use for public benefit, for registering land used for Party and government buildings and offices, municipal infrastructure, educational facilities, scientific research institutions, sports facilities, public culture, healthcare, and social welfare, etc. This should list everything in detail and be unambiguous. This way, government administrators or main market parties cannot take advantage of the situation in the name of public benefit to use land free-of-charge; buy it at cheap prices but then actually manage the land for profit; or undertake major construction projects that waste land.

3.4.3 There Should Be Substantial Rights for Disposition of Land

Now that rural land is collectively owned, do collectives have the right to transfer or develop land? No, they do not. According to Article No. 43 of Chapter 5 of the *Land Management Law*, any unit or individual who needs to use land for construction must by law apply to use state-owned land. One exception to this is for setting up village or township enterprises and construction of villagers' dwellings that are approved by law to use their own collective land. Another exception is for construction of public facilities or public welfare undertakings for villages (including towns) that are approved by law to use collective land of farmers. The state-owned land mentioned here that could be used via application by law includes land owned by the state and land that originally belonged to farmers' collectives but was then requisitioned by the government.

From this, one can see that collectives do not have the right to directly transact land with those that would use the land. Collective land must be transferred into government land before being used for construction purposes. The *Land Management Law* limits the rights of collective land to be directly offered for sale on the market. Legally speaking, in a universal sense, publicly owned property rights, privately owned property rights, collective property rights, and national property rights should all be equal. Public rights cannot violate private rights. In addition, national property rights cannot violate collective property rights. But the reality of agrarian law is that farmers' collectively owned land only has rights of possession, use, and usufruct, and does not have the rights of full disposition. It cannot change the nature of ownership rights by transaction and transfer, etc. Compared to state-owned land, collective land has no complete property rights and its rights are not equal with that of state land. To quote some local officials: [The *Land Management Law*] is a law that discriminates against collectives.

The fundamental reason that the rights of collective land are unequal to those of state-owned land is that public rights encroach on collective and private rights. It is also the main reason for numerous problems that occur in the process of land requisition. During a survey research, some local cadres jokingly claimed that land property rights of collectives and farmers were all so indefinite that the director of the land bureau could very quickly acquire farmers' lands. Collective land ought to legally enjoy the same equal property rights status as state-owned lands to make sure that both conform to the concept of "same land, same price, and same rights." Collectives and farmers should have the right to directly negotiate land transactions with land users. The land owned by rural collective economic organizations (RCEOs) should not only have rights of land ownership, usage, and usufruct, but should also have rights of disposition and rights to be directly negotiated in the land transaction marketplace. After entrusting collectives with equal land rights, the government can still regulate collectives' and farmers' land utilization and transactions through various legal strategies such as various programs, quota controls, etc. By giving them these rights, it would not only reduce administrative costs and increase economic efficiency, but it would also achieve regulation and rational utilization of land.

The *Decisions* of the Third Plenary Session of the 17th Central Committee of the CPC proposed to: "Gradually establish a unified market for land used for construction in both urban and rural areas. For rural collective lands to be used for commercial construction obtained by law, land usage rights must be transferred through a unified, tangible market, and in an open and standardized way. In this way they can enjoy the same rights and interests as state-owned land, if conforming to long-term plans." However, reform on this aspect has made no headway.

In thinking about revisions to the *Land Management Law*, collective land to be used for construction should be allowed on the market under the concept of "same land, same price, and same rights." It is reported that the *Land Management Law* has been included three times in revisions of legislative plans of the People's Congress. But because of persistent opposition from local governments and related departments, it has never been passed. I also participated in submission of suggestions for revision of this law.

The Third Plenary Session of the 18th Central Committee of the CPC once again proposed to: "Establish a unified market for both urban and rural areas for land to be used for construction. Under the precondition of conforming to plans and usage controls, permit land, presently under collective group management but that is desired for construction, to be sold, leased, or to enter the stock market, and permit such land to enter markets with rights and prices on par with state-owned land."

In 2015, there will be pilot reforms that allow collective land to be used for commercial construction to enter the market. I personally believe that the next step should be focused on opening up routes for collective land to enter the market. The feasible method for this would be direct transactions (including transactions, leasing, and becoming shareholders of equity investment, etc.) between the subject that would use the land and the subject that owns the land. The subject that would use the land would provide recompense to the subject that owns the land through purchasing it, paying rent for it, or issuing shares of equity investment in it.

The government would be primarily responsible for three things: establishing a trading platform and market; examining the land-use plan and usage control; and drafting financial taxation systems of collective land used for construction purposes and then collecting taxes on it. If the transaction market is established (including transaction rules) and the tax system and value added income-distribution approach are made clear, then the collective land to be used for construction directly entering the market (to be sold, leased, or have shares of equity investment) would pose no problems. That way, collective lands that are wanted by towns and cities for construction could also enter the market. The concept of "same land, same rights, and same price" would thus be realized. If government interests could be insured by taxes, governments at all levels would not be against it. If transactions were open through unified trading markets, it would display the true value of the land and safeguard the interests of all parties. If there were unified auditing standards, then they would conform to the long-term plan and usage control of land, as is done now. If there were a reasonable division of profits, the different parties involved would reach consensus. Finally, no matter what kind of collective land to be used for construction (whether commercial, public welfare, or housing sites), it could all be transferred to the market. There should not be too many limits placed on transactions, as long as organizations themselves mutually and voluntarily agree on a reasonable price for the land.

3.4.4 Rights of Collective Membership Should Be Delineated

Unlike the explicit stock rights in corporations, in rural collective economic organizations there exist problems of unclear membership rights. This lack of clarity is manifested in two areas: one is that collective members are always in a state of flux; the other is how to distinguish between members of collective economic organizations and ordinary residents in the rural communities. Are people counted as members who are newly added to the community through marriage, birth, or from moving in from outside to settle? What about those who die, move out, or divorce? Are they still to be considered members of the collective economic organization? How can these distinctive types of members enjoy collective land rights? These concrete questions are important in every village and to every farmer.

Currently, land requisition is common; collective capital, i.e., property, is adding value rapidly in developed areas and suburbs. Unclear membership rights easily lead to disputes over allocation and compensation. Large numbers of rural laborers have left their farms to seek manual work elsewhere; they settle down in other places together with their whole families. The place where they have membership rights in a collective and the place where they live are separate and different. Therefore, the rights of collective members and their enjoyment of social welfare services are not in synchrony, causing confusion and numerous contradictions for individuals and society.

However, we should base the present situation on those historical facts since 1956, i.e., the realization of collectives based on legal rules, regulations, and government policy such as the national land contract law and primary-level autonomous organization laws, etc. We should consider local rules and conventions in rural areas as well as social customs and habits, and thus further define and make unequivocal the membership rights of collective economic organizations and other detailed rights that members can enjoy (e.g., land, the collective's shares of stock, etc.).

A few years ago, some public servants in a city in Zhejiang Province moved their official registration back to their ancestral villages. Why? It was because village housesites that were allotted to them were valued in the hundreds of thousands of RMB and some even more than 1 million RMB!

It is obvious that rural land is quite valuable; therefore, it is extremely important to clearly define membership rights. In the present era, we should do the following: assure membership rights for individual farmers who leave their land to work and settle down elsewhere; reform the management structure of rural society; and separate farmers' identities as members of organizations from their identities as residents in the communities where they live.

3.4.5 The Market Position of the Rural Collective Economic Organizations Should Be Clarified

What is a rural collective economic organization? I am unable to explain it clearly. The Constitution, the *Real Right Law*, the *Land Management Law*, and the *Land Contract Law*, etc., make clear that the legal status of RCEOs entrust them to exercise collective ownership of rights, property, and natural resources such as land.

Nevertheless, what is incomprehensible is, up to now, our country does not have any laws, regulations, or government policy documents that set out definitions or explanations as to what a rural collective ownership organization is. It is as if this basic economic unit has no need to be explained and is already very clear. Actually, it is otherwise! The question that matters most is that RCEOs have neither concrete form nor status as a main force in the market.[34] At present, in rural areas, the concrete manifestations of the Party, government, and specialized cooperative organizations are all extremely clear; only that of rural area collective economic organizations are somehow never rendered clearly in law and government policy.

In practice, with the exception of Guangdong and Zhejiang provinces, Beijing, and a few other areas, village committees frequently exercise rights and functions on behalf of collective economic organizations. Nevertheless, village committees are autonomous organizations of villagers and differ from RCEOs. This muddles the distinction between these two kinds of organizations as to their nature, rights, and functions. Objectively, this puts the rural political economy in a state of mixing together politics with economics. Both of these after all are different in nature; politics and economics should not be commingled. It is very difficult for village committees to carry out effectively the economic activities of RCEOs. Furthermore, the members of these organizations and the villagers of the village committees are defined differently; but the executive bodies of organizations are also equated with the village committees. In addition, the collective organizations have no legal status, so they are not able to deal with economic activity nor can they open accounts in financial institutions such as credit unions, banks, etc., and thus in reality special economic organizations such as RCEOs have no market status and are therefore not able to carry out market activities. In short, rural collective economic organizations have no "identification card" that is recognized by markets. As was said by a grass-roots cadre in the Wenjiang district of Chengdu, Sichuan Province, "The rural collective economic organizations can't get things done."

At the same time, under some certain circumstances the law has a shortage of detailed regulations concerning land ownership by village groups, village collectives, and township collectives. Owing to their unclear legal basis, practical operations tend to be at random; they frequently overstep their authority, which leads to the land and property interests of village groups and farmers being violated.

Village Party organizations, village committees, and farmers' professional cooperatives all have laws that they must follow. Only the RCEOs still have no specialized legislation; for a long time they have had no identity. It is proposed that there be a *Rural Collective Economic Organization Law* in order to guide their work. This would allow these organizations to do the following in a better way: enhance both their development and the collective economy by entering the momentous market economy; safeguard their independence and autonomy; and

[34]In 2010, the author published, "*Guanyu zhiding 'nongcun jiti jingji zuzhi fa' de sikao*" ("*Thoughts on formulating 'laws for rural collective economic organizations*'") in *Nongye jingji wenti* (*Problems in Agricultural Economics*), which discusses this question in exhaustive detail.

adapt to current changes in the rural economy and society. This law should cover the following main points: legal status; concrete form and name; legal corporate status and institutional code numbers [used for identification purposes for setting up bank accounts, paying taxes, etc.]; registration certification; rights, functions, and obligations; management rules; qualifications and verification of membership; and, rights and responsibilities of members. We should draw on the experience of corporate management, transforming these organizations into modernized economic organizations that specialize in the exercise of land and other collective properties.

In 2015, the CPC Central Committee's *No. 1 Central Document* stated that, "There should be no delay in researching and making a draft of regulations for rural collective economic organizations." In the future, rural collective economic organizations will have "identity" and "status."

3.4.6 Income Distribution Regulations Need to Be Strengthened

Collective land income mainly includes land requisition compensation, rental income, etc. As to how the income from land should be distributed, the *Land Management Law* and the *Real Right Law* have nothing to say; there are only descriptions of principles in some related documents.[35] In addition to the old problems concerning the distribution of land requisition compensation, new problems have emerged concerning land income distribution (including income from collectives' land derived from rentals, sales, exchanges, and development of collective land for construction).

The income distribution mechanism has even more problems. This leads to each locality going its own way with no unifying principle, leaving room for rent-seeking[36] and corruption. To a certain degree, this has infringed on farmers' interests and triggered a large number of disputes over land benefits. For this

[35] In October, 2004, the *State Council Resolution on Deepening the Reform and Land Management* pointed out that: "The People's Governments of provinces, autonomous regions, and municipalities under the direct jurisdiction of the central government should, on the basis of the principle that land compensation should mainly be used to compensate farm households that have had their land requisitioned, formulate distribution methods for land compensation within the rural collective economic organizations." In November, 2004, the Ministry of Land and Resources issued the *Guiding Opinion on Refinement of the Land Requisition and Resettlement Compensation System*, indicating, "The distribution of land compensation should be in accordance with the principle that such compensation should mainly be used to compensate farm households that have had their land requisitioned. This compensation should be distributed rationally within the rural collective economic organizations. The detailed distribution methods should be formulated by the people's governments at the provincial level".

[36] Rent-seeking is the action of individuals or corporations that seek to increase their share of existing wealth by manipulating the social or political environment (e.g., by lobbying the government for loans, subsidies, grants, etc.) without creating new wealth.—trans.

reason, improving collective land income distribution should be an important part of the agenda for reform of the land system, laying down explicit rules in related legal regulations and documents and sorting out those regulations from top to bottom.

I came up with a formula I call a "tax and share system." The idea is that government would formulate regulations on taxing the gain in appreciation on collective lands, collect a certain tax fee based on that, and then all levels of government, collectives, and farmers could enjoy sharing in the profits on such land appreciation.

3.5 Farmers' Land Contract Management Rights: From Restrictions to Empowerment

How much right do farmers have to the contracted land allotted to them? This is both a general and a specialized question. The contracted land management rights of farmers are undergoing change, moving from the restriction of rights in the past to the endowment of rights in the present. The *Real Right Law* now makes it explicit that contracted management rights are usufructuary rights of farmers. This is a significant development. According to law, farmers' collectives have usufructuary rights to collectively owned land, i.e., contracted land management rights and the right to housesites on the land. The rights of possession, use, and benefit from the land that had been the collectives' rights were alienated in favor of farmers. The rights of contracted management of the land means that holders of these rights, such as farm households, have rights by law to possess, use, and benefit from cultivated land, woodlands, and grasslands that they contract to manage. The right of contracted management of land as a usufructuary right means that farmers no longer just contract and cultivate the land from collectives and have no clear rights; instead, they process rights to own, use, and benefit exclusively from the land that they contract to management. This establishes the legal basis for protecting farmers' contracted land rights and it implies that the land rights of farmers have equal status with national land requisition rights and collective land regulation rights.

The *Central Party Committee's Decisions on a Number of Important Issues Relating to Promoting the Development of Rural Reforms* that was issued at the Third Plenary Session of the 17th Central Committee of the CPC pointed out that we should refine contracted land management rights and safeguard by law the rights of farmers to possession, usage, and income of contracted land.

The *Decisions* of the Third Plenary Session of the 18th Central Committee of the CPC entrusted farmers with more rights: Adhere to rural collective ownership of land by law; safeguard farmers' land contract management rights; and develop and strengthen the collective economy. Stabilize rural land contract relationships and maintain them for the long run. Furthermore, under the premise of adhering to and refining the most strict arable land protection system, entrust farmers with rights to possess, use, benefit from, and transfer contracted land, as well as entrust them with the power of mortgage and guarantee of their contracted management rights. Allow farmers to become shareholders in equity investments and to develop agricultural industrialization management. Encourage contracted management rights to be transacted in an open market for specialized entities, family farms, farmer cooperatives, and other such farming enterprises. Finally, encourage development of various forms and scales of operations.[37]

3.5.1 Contracted Land Is no Longer Frequently Adjusted

The right of possession refers to the privilege to possess and control properties such as land; it derives from legal provisions. These provisions are not the same as possession; possession refers to actual control of land. The right of possession of contracted land management rights means that individual farmers, as members of a collective, naturally enjoy corresponding possession of ownership rights to the land. This is stipulated by law. Collectives can take back certain specific land that farmers have contracted. But they cannot deprive farmers of the right to contract collective land. This is because the right to contract a part of collective land is a farmer's right as a member of the collective. The right of ownership is the basis for the rights of use, income, and disposition.

The farmer's ownership rights of contracted land have always been unstable. At the beginning of the rural reforms in the 1980s, there were constant adjustments in the size of contracted land that farmers cultivated. Every few years, adjustments were made. During my childhood in the 1980s, our family's contracted land was adjusted three times. Such frequent adjustments negatively influence farmers' enthusiasm to increase production and to make long-term investments.

In order to avoid frequent changes to contracts of cultivated land and to prevent endless small adjustments to farmland management, on March 28, 1995, the *State Council Memorandum to the Ministry of Agriculture: Suggestions Relating to Stabilizing and Refining Land Contracts* proposed stabilizing contracted land management rights. It was proposed to "Keep the size of the contracted land area the same, no matter whether the number of persons in a family changes or not."

The *Land Management Law* (1998) prohibits extensive village-wide adjustments; it permits small-scale adjustments only after the agreement of two-thirds of village representatives followed by the approval of the county government and its agricultural department.

Nevertheless, as my 2011 survey of farmland transactions indicated, since the implementation of the second round of land contracts, 29.6% of farm household

---

[37]This sentence basically means to increase the size of farm operations in order to increase efficiency and reap the benefits of economies of scale.—trans.

contracted land underwent adjustments. But as soon as one problem is solved, others pop up: after the implementation of the land policy of "Keep the size of the contracted land area the same, no matter whether the number of persons in a family changes or not," new problems emerged.

Do the large number of rural babies born after 1998 (the starting point of the second round contracts) have contract rights? Should minors living at that time, quite a few of whom have at present already started families and established separate households, now be given contracted land? These and analogous questions still have no rational answers.

In our 2011 survey of farmland transactions, one interesting statistic was that 60% of households hoped that land could be adjusted every few years in accordance with demographic changes. My personal suggestion is that we take the population at some given time as the standard; there should be one final adjustment and then no more. In an era of stability, land should not be divided constantly; rather, after equalizing the land once and for all, there should no longer be land divisions that rely on national administrative power. Land resource allocation should then depend on market mechanisms and the individual situations of farmers.

In summation, following verification and issuance of land registrations in the last few years, farmers have more and more rights to contract land.

3.5.2 Contract Period Will Be Even Longer

The instability of rights of possession and use are also manifested in the contract period. The *Rural Land Contracts Law* stipulates that: the contract period for arable land shall be thirty years; for grasslands, thirty to fifty years; and for woodlands, thirty to seventy years. Second round contracts of arable land will expire in 2028; after they expire, how will the land be contracted? The *Decisions* of the Third Plenary Session of the 17th Central Committee of the CPC suggested that land contract management relationships remain stable, long term and with no change; but what were referred to were land contract relationships, not time limits on contracted land. Seven years have passed; there should be clear indications about how to carry out a policy that is "both long term and with no change" and how the next round of contracts should be managed. Some related reports propose that after the contract period is over, contracts be continued for seventy more years; in my view, it should be extended even longer. To reference the opinions of farmers, the survey of farm households in 2011 showed that 90% of farm households considered that farmland should be given to farmers for long-term use and 62.6% of farm households believed it would be appropriate if contract periods were in perpetuity.

The verification and registration of documents of contracted land management rights are the basis for stabilizing the rights of possession and usage and for safeguarding land and property rights of farmers. In 2013, the CPC Central Committee's *No. 1 Central Document* proposed that within five years the whole country should complete the work of document verification and land registration work. This should be regarded as an opportunity to standardize land transactions and land requisition, etc. The ownership rights and usage rights in contracted land management rights are moving from instability toward stabilization.

3.5.3 *Land Requisition Compensation May Fluctuate in Response to Market Conditions*

Bestowed as part of the main body of contacted land management rights, usufructuary rights refer to the rights to the use or enjoyment of contracted land. As the nation has continued to bestow additional contracted land management rights, usufructuary rights have also been enhanced.

The classification of income from contracted land has always lacked a satisfactory summarization. My personal opinion is that contracted-land profits should mainly derive from earnings from management production [i.e., farm production], farmland transactions, compensation for requisition of land, and dividends from equity investments. The most basic type of income from contracted land is that derived from management production. That is, the holders of the rights of land under contracted management should reap the profits from the land that they plow. Farm production was also the dominant source of profits during the early period of the rural reforms.

Following the gradual loosening up of legal rules, regulations, and government policies on transacting contracted land management rights, profits from such farmland transactions have become the main source of profit for contracted land. My 2011 survey on farmland transactions showed that the yearly average profit from farmland transactions was 277 RMB per *mu*; at that time, an area of about 200 million *mu* of farmland was transacted throughout the country. In recent years, both the area of transacted farmland and rents for farmland have increased in tandem, with rents increasing so much that they have become one of the main costs of farm production.

The *Decisions* of the Third Plenary Session of the 18th Central Committee of the CPC proposed giving farmers permission to buy shares in stocks of contracted management rights in order to develop industrialized management of agriculture. Compared with previous practice that only regarded equity investment as a form of farmland transaction, this was a very big step forward. It also means that from now on, buying into equity investments and sharing dividends will become a new source of income for contracted land.

Over the last two decades or more, the amount of collective land government has requisitioned has increased and compensation for it has become a source of income for farmers. But such compensation is merely a one-time gain. At present, land requisition compensation uses the value of agricultural output as the basis for calculation; in general, each *mu* of land is compensated at a range varying from 30,000 to 50,000 RMB. This is in sharp contrast to the land's market price per *mu* that can easily be as high as hundreds of thousands of RMB to over 1 million RMB.

It is widely believed that compensation rates for land requisition are set too low; for this reason, there are loud cries to increase them. Whether land compensation payments are determined based on the value of the land for agricultural production or are based on market value requires reasonable proportion. Otherwise, it is empty talk to speak of compensation mechanisms and ways to allocate appreciation of gains in land values. If the standards are set too low, farmers' interests cannot be guaranteed. But if they are set too high, it will cause the following affects: Land requisition costs will increase, as will that of infrastructure construction. Attracting merchants or investments will be negatively influenced (as enterprises probably will not be able to obtain land at zero or low cost). Pressure on government will increase. The urbanization process will slow down. Finally, GDP and economic growth will be negatively influenced. Nevertheless, in terms of stimulating more frugal and centralized use of land, the invisible hand of the marketplace would be more effective than the hand of the government.

We should refine the relative articles in the *Land Management Law*. Calculating land compensation using agricultural output value as the basis is insufficient; market value of land should be considered in the calculations. Both government and the people should share benefits from the rising price of land by allowing income derived from the appreciation in land value to be rationally allocated with the nation, local governments, and farmers. Only in this way will it reflect a transformation of strategic thinking embodied in the sentence, "Industry nurtures agriculture and cities support the rural areas."[38] Only if the *Land Management Law* establishes the concept that market value is the basis for land compensation can the problem of low standards of compensation be rectified; accordingly, *Regulations on Requisition of Collective Land* will probably be issued to make this adjustment in the standards for compensation.

3.5.4 Contracted Land Can Be Transacted Under Certain Conditions

It is neither easy to explain nor to understand the rights of disposition that are incorporated in the rights of contracted land management. According to the *Real Right Law*, the holder of the rights for contracted land management has the right to possess, use, and derive profit from the land, but has no right of disposal. But why are holders of these rights able to autonomously transact land? Is this not also a form of disposal? This profound and subtle conundrum is because what the holders transact is the contracted land management rights. To say it more accurately, what they transact are rights to manage the land, not the land itself. Hence, the right of disposition refers to the rights that holders have to dispose of contracted

[38]This quote is taken from the Fourth Plenary Session of the 16th Central Committee of the CPC, held in Beijing in September, 2004—trans.

management rights, but not to the right to dispose of the land itself. The right to dispose of collective land resides in farmers' collectives. (Examples of this would be awarding internal contracts and conducting external transactions). Disposition rights include transfer, mortgage, inheritance, and other forms. Strictly speaking, in transacting and mortgaging, contract rights and management rights diverge. The land contract holders either transact or mortgage the right of management.

The transaction rights found in contract management rights are an important embodiment of the rights of disposition. Since the rural reforms, following the trend of widening opportunities for employment of farmers and the migration of the rural population, transaction rights have expanded continuously.

In 1984, *No. 1 Central Document, the CPC Central Committee's Notification Regarding Rural Work of 1994*, proposed that the land contract period be extended to fifteen years; this "cracked open the door" for farmland transactions. For the first time, it permitted limited transaction of rural land. That is, land was allowed to be subcontracted, but only to members within the collective organization; leasing it out to others was forbidden.

In 1993, *No. 11 Central Document, the CPC Central Committee and the State Council's Number of Policy Measures of Regarding Current Agriculture and Rural Economic Development* proposed that when the fifteen-year period of the land contracts expire, it be extended for an additional thirty years. It would permit transfer of the right to use the land, with payment of a fee according to law. This implies that (other than subcontracting) leasing and other forms of transactions also comply with government policy. At that point, the policy "opened the door half-way" for farmland transactions.

The 2001, *the CPC Central Committee's Notice Concerning Doing a Good Job on Transactions of the Right to Use Farm Households' Contracted Land* for the first time systematized government policy on farmland transactions. It pointed out problems such as coerced transactions, harm to the interests of farmers, and changes of agricultural use of land to other usages. It proposed comprehensive policies to guide healthy development of transactions of rights to use rural land. Transactions must proceed on the basis of voluntary participation, compensation, and in accordance with the law. They should not seek to imitate the outward form as pretext for doing something nefarious; should not be assigned quotas to be fulfilled; should not be implemented forcibly; and should not cause the transfer of laborers when requisitioning farmers' contracted land. The Central Party Committee did not encourage business enterprises to lease or management large tracts of contracted land for the long term.

Passed in 2002 and implemented in 2003, the *Rural Land Contract Law* for the first time made the rural land transaction system legally clear. This marked a milestone in our country among the government policies, laws, and regulations concerning farmland transactions. In Chapter 2, Section 5, of this law, there are twelve articles governing the transaction of contracted land management rights. They stipulate in detail the various items concerning farmland transactions. One of them is to make the form of farmland transactions more concrete. "The rights of contracted land management through household contracts can be transferred according to law by subcontracting, leasing out, exchanges, assignment of rights, or other forms."

The 2008 Third Plenary Session of the 17th Central Committee's *Decisions on a Number of Major Questions Relating to Promoting Rural Development and Reform* proposed that in farmland transactions the government's obligation is to strengthen management and service and to establish a robust market for farmland transactions. Transactions should be done in accordance with three principles: the rule of law, voluntary participation, and compensation. The five forms of transactions are: "subcontracting, leasing, exchanging, assignment of rights, and joint-stock partnership."

The objective of transactions is to develop various forms of large-scale management. Entities meeting certain conditions, such as specialized major producers, family farms, and professional farmers' cooperatives, can develop operations of sufficient size to reap benefits of economies of scale. The essence of Article 3 was the following three items: "There must not be a change of collective ownership of land; contracted land rights must not be harmed; and interests of farmers must not be harmed." With this, government policies on transactions of rural contracted land management rights were fundamentally established. Subsequent government policies and definitions all regard these three items as guiding principles. Under the influence of different factors such as government policy guidelines and the increasing number of farmers migrating into big cities to work and settle down, the transaction of contracted land management rights has speeded up. At the end of 2013,340 million *mu* of the nation's contracted arable land was involved in transactions. This was 5.3 times more than that at the end of 2007; it reached 26% of the total portion of such land, which was greater than the 20.8% reached at the end of 2007.

However, there still exist several theoretical and practical misconceptions about farmland transactions. As for theory, people did not quite understand the government policy and specific meanings in their narrow sense, e.g., those land transfers that also included land used for construction purposes. There were misinterpretations and misunderstandings by both the general public and the media, especially around the time of the recent two Third Plenary Sessions (the 17th and the 18th).

As for practical misconceptions, these led to such phenomena as forced transactions against the will of farmers, violations of the transaction rights of farmers, and changes in the usage of farmland, etc. As a result, relevant administrative departments had to give repeated orders to restrain these behaviors. In 2012, I wrote and submitted to higher authorities an article, *Treat Business Enterprises that Seek to Lease Farmland on a Large-scale with Caution to Prevent New Land Annexations*. This induced the Ministry of Land and Resources to issue a plain-language bulletin called *Urgent Notice to Strictly Forbid Business Enterprises from Leasing Farmland and Changing Its Land Use to Non-Agricultural Construction Without Authorization.*

In February, 2014, the Ministry of Agriculture issued notifications that emphasized the need to always adhere to principles of "voluntary participation, compensation, and adherence to the law." This was in order to solve problems in several regions concerning transfer of rural contracted land management rights. These problems included: one-sided pursuit of large-scale transactions or a high proportion of transactions; relying on administrative fiat; and unnecessarily using extensive, long-term commercial capital to lease and manage the contracted land of farm households. The Ministry of Agriculture issued notifications that emphasized always adhering to the three principles of "voluntary participation, following the law, and compensation." There must be no harm done to the interests of farmers, no change of land usage, and no sabotage of comprehensive farm production capability. Actions must be taken that suit local circumstances, with a step-by-step approach that respects the aspirations of farmers. There must be no repeat of the problems seen in the Great Leap Forward, no coercion, and no arbitrary or confusing administrative orders.

At the end of 2014, the General Office of the CPC Central Committee and the General Office of the State Council printed and distributed *Suggestions on Guiding Orderly Transfer of Rural Land Management Rights in Order to Develop Agricultural Management on an Appropriate Scale*, which went a step further in regulating farmland transactions. For those who have further interest in this topic, I refer you to *Zhongguo nongdi liuzhuan wenti diaocha* (*Investigation into farmland transaction issues in China*), a book of which I was the lead author.

3.5.5 In the Future, Management Rights Can Be Mortgaged

The right to mortgage contracted land management rights has always been strictly limited. The *Mortgage Law* and the *Real*.

Right Law[39] strictly forbid farm families to pledge their rural land management rights to obtain mortgage loans. However, this is a controversial topic and subject of heated debate. Those who approve of this kind of mortgage consider that mortgage rights should be one of the rights included in contracted land management rights, as

[39]The *Real Right Law* permits mortgaging of contracted land that has been acquired by other methods. According to its Article 133, by way of inviting tenders, auctioning, and opening negotiations on contracting wasteland and other such rural land, and in accordance with statutes and related pertinent regulations of the State Council, the rights pertaining to such land includes the ability to transfer, lease, offer equity investment, mortgage, or the use of other methods of transactions. A draft of the *Real Right Law* almost permitted mortgaging of contracted land management rights. According to a report of the Xinhua News Agency on October 27, 2006, the sixth review of the draft of the Real Right Law deleted related regulations from the previous fifth review that would have permitted, under certain conditions, the holders of contracted land management rights to mortgage those rights. The draft of the fifth review stipulated that the holders of contracted land management rights who had stable sources of income could mortgage their rights with the consent of the party who had awarded them their management contract. Those who obtained such rights to mortgage would not have been able to change the usage of the contracted land.

it would alleviate the problem of farmers having difficulty obtaining loans. Those who oppose it believe that if farmers lose their land because of not being able to repay mortgage loans then this would precipitate social problems.

In 2010, we conducted a survey in Tongxin County, in the Ningxia Hui Autonomous Region, and wrote a report called *Mortgaging Rural Land Management Rights Is Worth Promoting*. The key idea in the report was that in the mortgage mode that they practiced there, the farmers only mortgaged their contracted land management rights for a certain period of time; because of this provision, they would not lose their land. The Tongxin experiment made it clear that it was feasible for farm households to mortgage their rights of contracted land management and that it was worth experimenting and spreading the concept. The law and government policy should permit these kinds of mortgages. After the survey report was published, it was criticized, which left us feeling gloomy.

When I made my 2011 farmland transactions questionnaire for farm households, I especially designed questions to look at farmers' views on mortgaging land; the results revealed that 67.6% of farm households agreed that farmers should be allowed to mortgage their land.

The *Decisions* of the Third Plenary Session of the 18th Central Committee of the CPC made this clear by giving farmers the right to use contracted land management rights to secure mortgages. This legalized those informal experiments on this issue that were occurring in several areas and it also gave a green light for farmers to exercise mortgage rights in the future.

3.5.6 Inheritance Rights Should Be Clear

Inheritance rights of contracted land management rights have never been made clear. The law provides no clear authorization for contracted land management rights to be inherited from generation to generation. This can be seen from the regulations in the *Land Management Law*, where it stipulates that contracted rural land is a kind of contract action but with no regulations regarding land inheritance. The *Real Right Law* does not say a word about the question of whether rural contracted land can be inherited. The *Rural Land Contract Law* separates land contracts into two categories: family contracts and other manner of contracts. According to its Article 31, "Contract profits earned by contractors shall be inherited in accordance with inheritance law. When the contractor of a woodland dies, the contractor's successors can continue the contract within the contract period." Article 50 states that, "When a contractor dies holding land contracted management rights obtained by bids, auctions, or public agreements, etc., then the earnings due under the terms of the contract can be continued according to inheritance laws; within the contract period, the heirs of the deceased can continue the contract." There is no use of the term "inherit" in illustrating the concept, but rather it uses the expression "continue the contract." This serves the purpose of encouraging management of wasteland, but it does not touch upon household contracted land.

The *Decisions* of the Third Plenary Session of the 18th Central Committee of the CPC brought the idea of giving farmers inheritance rights for shares in the capital assets of collectives, but still was not clear about inheritance rights of contracted land management rights, leaving space for further adjustment in the future.

I propose that there ought to be authorization for the inheritance of contracted land. I suggest changing Article 31 of the *Rural Land Contract Law* into: "Contract profits earned by contractors shall be inherited in accordance with inheritance laws. When a contractor dies, successors can continue the contract within the contract period." Articles relating to conditions and limitations of inheritance of rural contracted land management rights should be added. Corresponding articles should be added to the *Inheritance Law*.

3.5.7 Methods for Compensated Exits Are Still Being Explored

The right to paid exit from land contracted management rights is a big topic that is novel, complex, and controversial. Farmers hold various opinions about this; in the 2011 survey of farmland transactions, 51.6% of farm households believed that after farmers go to cities to settle down, contracted land ought to be permanently retained by them; the rest of them felt that such contracted land should be returned. For a one-time payment of the sum of 88,000 RMB per *mu* of land, farm households would be willing to be bought out or withdraw from the contracted land.

The *Decisions* of the Third Plenary Session of the 18th Central Committee of the CPC proposed giving farmers rights to paid withdrawal from shares of equity investment. However, it was unclear whether this included shares in land assets. According to the *Land Contract Law*, after farm households enter the established districts of a city, they should return their contracted land without compensation; this stipulation is no longer suitable due to urbanization and the integration of urban and rural areas.[40]

Are farmers able to carry with them their land ownership rights when they become city residents? Are they able to liquidate their land and property for a reasonable consideration before moving to the city? In the process of farmers turning into city residents, how are lands and collective property of farmers to be appropriately disposed of? How are farmers' land rights to be safeguarded? How are rural collective's economic property rights to be formed in order to further deepen and move forward the process of integrating urban and rural areas? How is

[40]Article 26 of the *Rural Land Contract Law* states that within the contract period, when a rural household moves to an established district of a city, the registration of the members of the farm household changes to that of non-farming residences and their farmland and grassland should be returned to the party that had originally awarded the contract for it. If the household that took out the contract does not return the property, the party awarding the contract can take back the contracted farmland or grassland.

the positive expansion of the city and orderly withdrawal of rural areas to be made complementary so that they prosper together? How are the completely different managerial systems of the urban and rural areas to be integrated? Governments at all levels, especially urban governments, must seriously face these vital issues.

In many suburban districts, a popular practice is that after requisitioning rural land and giving farmers compensation, the official identities of these former farmers are then converted into those of city residents. However, in the majority of cases, this transformation of official identity is in name only; these farmers-turned-city-residents are not treated at all the same as existing urban residents in respect to medical insurance, retirement benefits, etc.; the disparity in their treatment is great.

More importantly, numerous landless farmers, after changing into city residents, have lost the land that they formerly relied on to live and in effect have lost their land ownership rights. Other than receiving a little compensation, they have obtained a document attesting that they have city resident status, but nothing more.

For these reasons, it is unfair to ask that farmers abandon their share in collective land rights in exchange for receiving city residence status and to lose their land for only trifling compensation. In the process of farmers becoming city residents, we should enhance safeguards of farmers' land and property rights. With regard to migrant farm workers moving to cities to settle down and work, we should guarantee their land rights in the countryside; there must be no eroding of their contracted land management rights, housesite rights, and other land rights. We should allow, encourage, and regulate farmers who come to the city to work and live as city residents while letting them retain their land rights. With regard to those farmers who have already lived and worked in the cities and wish to withdraw from their land rights, ways should be explored to allow them to be paid to relinquish these rights. Using rural property rights transaction markets as a platform and offering reasonable prices for land property through these markets will both safeguard the land property rights of farmers and gradually improve efficient utilization of key aspects of the land.

Chapter 4
Can Farming Make Money?

Last year there was a flood, so the harvest was bad and we lost money. This year was considered a good year with a bumper harvest. But we still lost money!

—from *Duo shou le san-wu dou* (*This harvest is greater by three to five pecks*), by Ye Shengtao (1894–1988) (This is a short story that Ye Shengtao, a noted writer, editor, publisher, and educator, wrote in 1933.—trans.)

That was what a farmer said in Ye Shengtao's short story, after the farmers went to the wholesale grain shop and sold all their grain from the mid-autumn harvest. It reflects the reality for farmers—in a bad season, there is crop failure, yet in a good season, cheap grain hurts farmers. It reveals the problem of "High risks" and "Low profits."

As a farmer who worked on a small farm from a young age and as a researcher who has studied government agricultural policies for more than ten years, I have mixed feelings of love and pity toward farming. As far as love is concerned, there is the sweet smelling soil and wafting fragrance of the ripened grain. Ever since I was small, I grew up eating the millet, corn, sweet potatoes, potatoes, and pumpkins that my mother and father raised.

As for pitying the farmer's lot, it is hard to explain in a few words. Rooting in the soil for food is an extremely difficult way to make a living. There is no guarantee of a successful season when tilling the land. What is most feared is that after sowing the seeds there will be no rain. Then there is the thinning out of the seedlings and weeding under the blistering sun that must be done. There is the harvesting and curing the harvest, while fearing rain that could ruin it. There is storing the harvest, which must be guarded against insects, mice and rats, and the damp. Finally, at the end of a year of farming, one has just enough food and clothing to make do and nothing more. In a self-sufficient, traditional farming society, it was enough just to eat and drink one's fill. Farming means beginning work when the sun comes up and retiring only when the sun goes down; it is the

Y. Zhang, *Insights into Chinese Agriculture*,
https://doi.org/10.1007/978-981-13-1050-8_4

accepted way of life in the countryside. Nonetheless, in the current society of commercial goods and industry, with the cities and towns as the centers of modern society, farm profits are thin and risks are great. Apart from providing agricultural products, what chance do farming and farmers have for a prosperous future?

4.1 Natural Disasters Occurred Frequently

4.1.1 Since Ancient Times, Farming Has Been at the Mercy of Heaven

Good rain knows the correct season to fall, so in spring it occurs just in time to help seeds sprout.
Following behind the wind, it sneaks into the night and moistens all living things, delicately and without a sound.[1]

This joyous blessing only occurs in places of South China such as the Chengdu Plains of Sichuan Province's "Land of Abundance."

However, my native place is the mountainous district of Lüliang County in Shanxi Province, with its steep slopes of loess, where there are "nine years of drought out of every ten years." Every spring, the rains there are as precious as edible oil. The farmers in the mountainous areas grow crops completely dependent on the weather; if there is a spring without much rain then it spells trouble. Sowing crops then has to be changed from spring to summer. If the face of Heaven smiles on us with rain, then we cannot miss the opportunity to sow. However, it is still necessary to forecast the weather. Like chefs trying to measure food temperatures, farmers try to understand weather. Sowing crops a few days before it rains is the best; the seeds will germinate quickly and well. Re-sowing after a rain is far from ideal because right after the rain the ground is too damp to sow well. If one waits for the land to dry and then begins to sow, it would possibly delay the farming season. Furthermore, one still does not know when the next rain will fall. The worst possible scenario would be to have no rain after sowing the seeds, and then have to replant later.

My father was a farmer who understood the weather. Each year, spring planting at our house was always just right. We did not waste energy nor squander anything, and if the growth and harvest were not bad, we would inevitably garner the envy of our fellow villagers.

[1]This is an excerpt from *Chunye xiyu* (*Welcome rain on a spring night*), by the Tang poet Du Fu (712–770), a poem so widely known in China that it needs no citation for the Chinese reader.—trans.

According to my observations and analysis (as a nonprofessional farmer), the weather seems to be changing in favor of agriculture in the north; in recent years, the rains in my home village in Shanxi have been heavier than in the past. Winters in the north have not been as cold as in my childhood. The frost-free periods have become longer and cumulative temperatures have risen. Could this perhaps be a slightly favorable aspect of the unfortunate greenhouse effect that is harming the earth's ecology?

Appreciation of Poems about Agriculture

Xia ri tan (Summer sigh)

—Excerpt from a poem by Du Fu, Tang dynasty poet

… Heaven has long withheld its thunder,
Is this not a most perverse heavenly command?
No rain moistens the living things,
Fertile fields rise into clouds of yellow dust.
Soaring birds die from searing heat,
Fish dry up as ponds turn to mud.
A myriad of people wander about, destitute and homeless.
Lifting up one's eyes reveals only a plethora of weeds….

This poem depicts the great drought and famine of 759 in the central plains of the Wei River basin in present-day Shaanxi Province.

From times immemorial, farmers have suffered from natural disasters. In the long sweep of history, when knowledge of natural science and technology was scarce, farmers could only submit to the will of Heaven. Consequently, both among the common people and in the courts of the rulers and ministers of successive dynasties, heaven and earth were worshipped. Imploring favorable weather for crops and an abundant harvest of grain became an important ceremonial rite and tradition. At the Temple of Heaven, one of Beijing's architectural landmarks, the Hall of Prayer for Good Harvests was originally called the Hall of Prayer for Grain. It was precisely there that on the first month of the lunar year the emperors of the Ming and Qing dynasties came to pay respect to and beseech Heaven for a bountiful grain harvest. Its circular design represents Heaven and the twenty-eight supporting pillars made of Chinese cedar, arranged in three layers from the inner to the outer, represented one year with its four seasons, the twelve months, and the twelve two-hour periods of the day, respectively. In total, they represented the essence of ancient astrology and astronomy, the twenty-eight Lunar Mansions, which are roughly equivalent to constellations or star groupings.

Appreciation of Poems about Agriculture

Chun Yu (Spring rain)

—by Xie Jin (1369–1415), noted Ming dynasty scholar

Spring rain is as precious as edible oil; when it rains, it flows, filling the streets.
Slipping down on the slick ground discomfits the scholar, who is then mischievously laughed at by passersby – they are such a herd of cows.

In this scene, the poet depicts the great value of spring rain, the poet slipping down, the passers-by breaking into laughter at the sight in a comedic and mischievous manner, and the poet in turn mocking them.

4.1.2 Natural Risks Persist to This Day

In addition to droughts, there is also damage to crops caused by flooding, heavy snows, strong winds, hail, and calamities of plant diseases and insect pests, etc. These are all natural disasters that agriculture has no alternative but to face. As I understand it, natural agricultural calamities can be divided into two large groups: those that are caused by weather and those that are of biological origin. Weather-related calamities include droughts and floods, etc., while biological calamities include epidemic plant diseases, major insect infestations, and livestock epidemics.

In recent years, because of factors such as the greenhouse effect and ecological damage, climate calamities have become more severe around the globe, including in China. Extreme weather has occurred frequently; natural risks have become more evident. Droughts and flood disasters that used to occur once every few decades now occur from time to time. The scope, frequency, and severity of droughts have increased; other disasters such as floods and frost damage have worsened. At the same time, climate disasters and disasters due to crop diseases and insect damage are interwoven; in sum, they have caused severe damage to crops.

Under the influence of extensive, centralized stockbreeding, risk of livestock epidemics has become a large potential problem. Epidemic diseases such as foot and mouth disease, porcine reproductive and respiratory syndrome, avian influenza, etc., are hanging over the head of the livestock breeding industry like the sword of Damocles.

"If you have money, don't waste it on feeding creatures with mouths [e.g., cows, hogs, chickens, ducks, etc.]." This was my father's viewpoint. So in the decade of the 1990s, father organized several fellow villagers to raise longhaired rabbits; but because of epidemics that caused the death of many of the rabbits, he lost money. Then several years ago he teamed up with others to raise silk worms, but this enterprise also suffered financial losses due to epidemic diseases.

Farming really is risky and investment must be prudent.

4.2 Market Risks Are Increasing

The risks of the agricultural market are reflected largely in price fluctuations and products that are "difficult to sell." The range and frequency of fluctuations in farm commodity prices are expanding and producers and business operators are facing greater and greater economic risks because of the quickening process of marketization and commercialization; increasing influence on agriculture of global markets; and the transformation of farm com-modities into objects of capital speculation. The related problem of "difficult to sell" has been constantly increasing in recent years. This is manifested particularly in cash crops such as vegetables and fruit; it has caused huge losses to farming operations.

4.2.1 Economic Principle Behind "Cheap Grain Hurts Farmers"

In *This Harvest is Greater by Three to Five Pecks*, the farmers harvest three to five pecks more per *mu* than in the previous harvest. This increased production was supposed to be a good thing, but when the supply of rice on the market greatly increased, prices fell; since the fall in prices was greater than the increase in production, the income from the sale of rice was less than that of the previous year. This is exactly what is meant by the economic phenomenon expressed by the short story's "Cheap grain hurts farmers."[2] It is certainly not simply a matter of the greed of the rice merchants. "Last year, when rice was expensive, the soldiers' rations were cut; this year rice is cheap, so farmers are severely harmed."[3] Western economists use the model of supply and demand and the theory of price elasticity of demand (PED) to explain scientifically the economic phenomenon of "Cheap grain hurts farmers."

Price fluctuations in farm commodities, especially necessities such as rice and wheat, have a relatively small influence on people's demands; people do not really consume one more *jin* of rice just because the price of price is reduced or vice versa, which means the PED of farm products is small. In a year with a bumper harvest, rice production increases, supply and demand fluctuate slightly, and the price falls. Because the PED of rice is small, the degree of price decline is greater than that of the increase in production. Multiplying these two factors, yield and price, results in a decline of income. This is also the economic reason for the national policy of providing a minimum price support for grain [i.e., a standing offer to purchase all grain at a set minimum price]. If grain were completely marketized [i.e., with no

---

[2]This phrase comes originally from the *Hanshu* (*History of the* [*Former*] *Han Dynasty*), *Zhaodi ji*. —trans.

[3]This phrase is from *Suiyan xing* (*An end of year dirge*), by Du Fu, translated below.—trans.

such minimum price support], farmers would be like a small skiff being tossed in the midst of a raging storm.

Accompanying the problem of "cheap grain hurts farmers," there is the problem of agricultural products that are "difficult to sell." These twin problems are also aggravated by other reasons, such as the seasonality of farm production and difficulties of storage and distribution of farm commodities. During the harvest season, commodities, such as vegetables and fruit that are not easy to store and must be sold as soon as possible, flood the market. At the same time, farmers want to recycle capital in order to reinvest. Adding to this tension is the problem of large supplies; all of these factors tend to force down prices, thus producing the "difficult to sell" phenomenon. Every year, there is no dearth of farmers who have taken great pains to plant vegetables and tend fruit but who, because prices are so low that they cannot find a market, must let their produce rot in the field. This distresses both those farmers and those who read reports of it. Virtually every year this situation occurs and the cycle goes round and round. It only varies as to which products are affected, e.g., vegetables, fruit, milk, etc.

Appreciation of Poems about Agriculture

Suiyan xing (An end of year dirge)[4]

—Du Fu, Tang poet

At the end of the year, the north wind blows everywhere;
The Xiaoxiang River and Dongting Lake [in Hunan] are covered in a flurry of pure-white snow.
The cold weather freezes the nets of the old fisherman,
So the Moyao tribesmen have no alternative but to shoot with a twanging sound at the wild geese, using their bows made of mulberry wood.
Last year, when rice was expensive, the soldiers' rations were cut;
This year rice is cheap, so farmers are severely harmed.
High officials, with great carriages pulled by teams of horses, weary of battening on wine and meat,
While the people are exhausted from working at their looms in bare thatched huts.
The Chu people[5] *prefer to eat fish and not birds—*
You must cease the wanton slaughter of the swans flying south.[6]
What is more, I have heard that everywhere sons and daughters are being sold.
In order to pay for the taxes of grain and corvée that their parents owe,
They will be sold, never to be reunited with their mothers and fathers.

[4]Based on records of Du Fu's itineraries, this poem can be dated to the end of the year 768, in the aftermath of the disastrous An-Shi Rebellion of 755–763.—trans.

[5]The Chu Kingdom roughly encompassed what is now Hunan Province.—trans.

[6]This line not only expresses the poet's sympathy for the swans and geese, but also by association calls to the reader's mind the expression *aihong bianye*, an idiom that literally means, "everywhere there is the plaintive whine of geese," and figuratively, "a land that is swarming with wailing disaster victims."—trans.

In the past, it was strictly forbidden to privately mint coins;
But now, the debasing of copper coins with too much lead and tin is tolerated by officials.
The easiest coins to obtain are shoddy ones cast from molds carved of clay.
Mingling of good and bad coins must not be allowed to continue to deceive people.
The painted military bugles of war blow from atop city walls everywhere;
When shall their plaintive and aggrieved tunes end?

This poem portrays the social and economic chaos and the bitter life of the people in the aftermath of the Tang dynasty's An-Shi Rebellion. In his poem, Du Fu writes about economic pheno-mena, such as how farmers are harmed by both high prices and low prices for grain and how the illegal minting of coins causes bad money to drive out good money. It would seem that Du Fu was not only a great poet but was also an early economist.

4.2.2 Explanation of the Hog Cycle and the Cobweb Theorem

In the first quarter of 2014, the live hog market was in a slump. The price of slaughter hogs in Sichuan and Shaanxi provinces fell below 6 RMB per *jin*; this was lower than the hog breeders' break-even parity price point[7] of a hog-to-corn ratio of 6:1.[8] For every hog, the breeders lost over 100 RMB.

The "hog cycle," which reflects the rise and fall of prices and supply fluctuations on the live hog markets, once again entered its nadir. The general pattern of the hog cycle is popularly expressed: "As the price of pork drops—breeders reduce the number of farrowing sows—the supply of live hogs then decreases—pork prices then rise—breeders then increase the number of farrowing sows—the hog supply then increases—and pork prices then tumble."

The live hog industry and market has gone through this cycle again and again, except that the length of the cycles is not always the same. It almost becomes a predictable pattern of occurrence. Breeders who find themselves in this situation suffer greatly from dramatic fluctuations of prices.

[7]For farm commodities, the parity price is the purchasing power of a particular commodity, relative to a farmer's expenses (e.g., wages, land rent, equipment, etc.) to produce it. The concept is that the selling price of produce should go up in the same ratio as the prices of the inputs used in its production.—trans.

[8]In China, the hog-corn ratio equals the live-weight hog price per kilogram divided by the feed corn price per kilogram. All other things being equal, any ratio higher that 6:1 would have made money for the farmer; any ratio lower than that would have lost money.—trans.

Pork prices are like riding a roller coaster

The cobweb theorem in economics (which was proposed by several Western economists in the 1930s) explains the pattern of the hog cycle. It refers to the theory that changes in the price and production volume of commodities mutually influence each other, which then gives rise to regular cycles. Its moniker comes from the fact that a diagram of supply and demand, prices, and production volume show changes that are interconnected in ways that resemble the strands of a cobweb. To apply this to the hog cycle, breeders in general decide their current breeding activity based on past pork prices, or their future breeding based on current prices. Ergo, when those prices fall, they have no choice but to sell at a loss. (If they do not sell, then their losses become greater for each day that they have to continue to feed their stock.)

They either quit the business or reduce the number of hogs that they breed. When numerous breeders respond in this way, market supply shrinks and prices then increase. When the prices again reach a level of profitability, breeders once more jump into the market one after the other, increasing the number of hogs bred. When the market supply increases to a certain level, prices drop… Breeders frequently misstep or drift with the tide, just as the majority of people do who buy stocks. They in effect buy high and sell low instead of the other way around, with resultant heavy losses.

Farmers have already suffered greatly from the hog cycle. But to make matters worse, in the past several years excessive production has spread extensively within the hog industry as a whole: non-agricultural enterprises such as steel, Internet businesses, etc., have also entered the hog breeding business, creating severe overproduction. To date, data from the Ministry of Agriculture shows that the number of sows on hand has reached 47.91 million head. This far surpasses a rational level of 25 million head. Capital from industry, commerce, and large breeding farms has solid power and its ability to resist economic storms is strong, causing ordinary breeders to be even more squeezed by the hog cycle.

In contrast to the drop in pork prices, the price of fresh ginger rose at the beginning of 2014, with its retail price reaching around 8 RMB per *jin* and with some districts even exceeding 10 RMB. This already exceeds the retail price of pork, which is around 8 RMB per *jin*. In the past few years, ginger, garlic, mung beans, and other such small kinds of agricultural products have experienced roller coaster prices. (The media even has popularized jokes and clever puns about ginger, garlic, and mung beans to make light of the situation.) These examples have many of the features of "diverging cobwebs" in the cobweb theorem.

In addition, these small items are prone to speculation; investment and speculative capital have added fuel to the fire. They have accelerated the imbalance between market supply, demand, and prices. When prices rise, investors and speculators make a fortune; when prices are in a slump, farmers lose money.

4.3 Profits Are Squeezed from Both Sides

4.3.1 Farming Costs Continue to Rise

Since the 21st century, Chinese agriculture has entered an era of "High costs" and "Low returns." In the long-term, with the gradual increase in the prices of primary products such as petroleum, the rural labor force will see an outward migration and labor costs will increase. Furthermore, transactions of farmland and rent for farmland are increasing. Therefore, the "High cost" is becoming more and more prominent. In the eight-year period from 2006 to 2013, the total cost for each *mu* of rice, wheat, and corn has risen from 445 RMB to 1,026 RMB, an increase of 1.3 times, 47% higher than the increased rate of output. The speed of increase of costs has exceeded that of the increase of output value, which has compressed room for profit from grain production. During these eight years, the average net profit per *mu* of these three grains plunged from 155 RMB to 73 RMB. Among them, the profits from rice were slightly higher. In 2013, the average net profit per *mu* of rice was 155 RMB, corn was 78 RMB, yet wheat lost 13 RMB; but the average net profit per *mu* of soybeans and cotton was 34 RMB and minus 215 RMB, respectively.

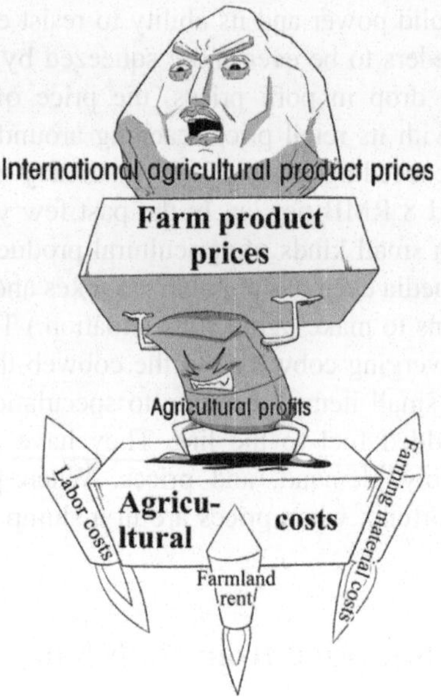

Agricultural profits are squeezed from both sides

The cost of agricultural production is an important integral part of the total cost of farm products. In recent years the rate of increase of this cost has been comparatively large, directly pushing higher the cost of production. Since the beginning of the new century, the price index of production has steadily risen; this was especially true in 2008, when it increased 20% over the previous year. From 2000 until 2013, the total price of agricultural production increased 90%. During the same period, the price of chemical fertilizer nearly doubled and that of animal feed more than doubled. In the last two years, under the influence of lower prices for petroleum and a sluggish global economy, the cost of agricultural production, mainly the price of petroleum-based chemical fertilizer, has decreased slightly. However, land rent and labor costs have gone up. The gradual rise in the cost of agricultural production has squeezed the room for profit, dampening the enthusiasm of farmers to produce, pushing up the cost of agricultural products, and adding to the burden of consumers.

4.3.2 *Difficult to Increase Price of Grain: "The Basis for the Cost of All Other Things"*

Agriculture, especially the growing of staple foods, is a low profit industry; consequently, the income of farmers is also low. Someone will ask, "Well, if we raise prices a bit on grain and other agricultural product then won't farmers income correspondingly go up?" It is okay to do such simple math, but the problem is that the price of grain has always been called "The Basis for the Cost of All Other Things."

This means that increasing grain prices will simply push the prices of other farm products higher, simultaneously pushing up the prices of other commodities; the result will be to increase the Consumer Price Index (CPI) and accelerate the rate of inflation. If the CPI exceeds a fixed mathematical value (3, 5%, or even higher), it then will increase the living expenses of ordinary people too much, especially those who have low incomes. It can also bring about other economic and social problems so that government has no alternative but to get involved. For this reason, government adopts certain measures to control the domestic price of principal agricultural products, including grain. For example, it offers guidance on grain prices and sells grain reserves when grain prices rise, so that the prices of grain and other farm products remain stable. This is unlike the situation in Japan and Korea.

The prices of agricultural commodities in Japan are conservatively figured at five times more than similar products in China, but there are only 2.5 million farm households in Japan. The working population in the primary industry of agriculture only occupies 5%. Therefore, Japan is able to carry out more supportive measures in the areas of price, credit, infrastructure construction, etc., thus safeguarding their agriculture and farmers' income.

But our country has 260 million farm households, with a working population in this primary industry numbering also about 260 million. [The reason that these two figures are the same (260 million) is that vast numbers of farmers have migrated to urban areas to work.] This number is so enormous that the nation, related secondary and tertiary industries, cities and towns, all find it difficult to effectively nurture and repay agriculture and farmers for their essential support.

4.4 Roads to Profit Are Diverse

In recent years, quite a number of non-agricultural enterprises and industrial and commercial capital have moved into the business of agriculture. Putting aside those that attempt to profit by enclosing and occupying rural land, are the farsighted ones not smelling the sweet aroma of profits in farming?

4.4.1 Hog Breeders' "Opulent Banquet"—Delicious-Looking but Poor-Tasting

More than 60% of the meat consumed in China is pork; pork possesses a huge market. As part of that, the pig-breeding industry is also a large, intensive agricultural industry. For this reason, it has been attracting all sorts of capital. Famous enterprises entering the hog-breeding business include NetEase, an Internet technology company; Wuhan Iron and Steel (Group) Company; Shanxi Coking Coal Group, etc. In addition, there have been many other, far less well-known enterprises, both large and small, that have entered the hog business. Relying on their ample supplies of capital, these business enterprises can realize economies of scale and standardized breeding of pigs, bringing an industrial model to the hog industry.

However, entering the hog business is no easy task. These enterprises must face pragmatic problems such as constructing factory farms, selective breeding, dealing with hog waste, and fluctuating or sluggish prices. After a five-year start-up period, the hogs raised by NetEase began finally hitting the market in 2014; according to reports, the first batch only numbered several hundred head. A considerable number of hogs have been raised by other enterprises, but the pork market in recent years has been in the doldrums. The live hog industry is seeing difficult days.

In my opinion, in keeping with the pattern of the "hog cycle," the main reason for the relatively long period of stagnation of the live hog market is that business enterprises coming into the hog breeding industry in a big way have created excessive production. With higher feed prices, such as domestically grown corn compared to cheaper foreign imports, and overproduction in the hog industry, it seems that the tempting "opulent banquet" of hog raising has turned out to be not that tasty.

4.4.2 High Quality, Safe Agriculture Has a Sunny Future

Although the proceeding summary analysis of farmers' profits has been pessimistic, agricultural products are still the most important necessities for people; agriculture is absolutely the most enduring and vital of industries. Obviously, it is not like some industries or products that are mere flashes in the pan. In the wake of people's need for varieties of food and higher demands for food quality and safety, one can predict that high-quality, safe agriculture is still an industry that can expect to be highly prosperous and profitable.

In recent years, many investment companies and commercial enterprises have expressed intense interest in this area; what is more, some have already started dabbling in it and have become "The New Farmers." To name some examples appearing in 2014, there are famous brands such as "Chu Oranges," "Liu Peaches," "Pan Apples," "360 Anxin Rice," Alibaba's "Ju tudi," and the high-end farm products of the Hengda Corporation. The parent companies of these agricultural brands were never before involved in farming. Although some of their businesses may not yet be running smoothly, nonetheless they still consider that high-quality, safe agricultural products will be both welcomed by consumers and profitable. Indeed, such products do have a sunny future. These enterprises utilize their own advantages, such as capital, reputation, market influence, and e-commerce platforms, etc., enriching what agriculture means. Nevertheless, it is important to remember that those making these kinds of investments must be prudent rather than simply blindly following trends.

4.4.3 E-commerce—Boosting Profits

The foundation of agricultural profits is sure to be agricultural products that are safe and high quality. However, "deeply buried gold cannot shine."

Seeing the rider galloping toward the city in a cloud of red dust, the imperial concubine smiled.
No one else knew that this was because the rider was bringing her lychees from afar.[9]

In ancient times, transportation and logistics were inconvenient and the trade volume of agricultural commodities between districts was small. Only a person with the wealth and status of a Yang Guifei could have relays of fast horses deliver fresh lychees from the distant south.

Nowadays modern transportation and logistics are convenient and quick, and even ordinary people in the north can enjoy easy access to fruit from the south. At the same time, southern fruit markets have expanded, selling even more widely afield, which inevitably increases profitability. Modern e-commerce, logistics, cooperative organizations, business methods, management, and marketing, together with investments methods, have all led to agriculture displaying new vitality and potential for profits. These are "wings" that have lifted up modern agricultural profits.

In 2013, when I took a temporary position as assistant mayor of Guangyuan, a county-level city in Sichuan Province, the story of two farmers who opportunely used such modern "wings" left me with a deep impression. One was a female undergraduate student from Qingchuan County in Guangyuan City who was born in the 1980s. After being graduated, she returned to her hometown to open an online shop selling local specialties such as edible black fungus. Utilizing modern e-commerce to sell high-quality local specialty products from deep in the mountains has led to great success for her. At the same time, top-quality, safe farm products have thus had the opportunity to enter distant markets.

[9]These lines are excerpted from the well-known poem, *Guo Huaqinggong, jueju san shou* (*On passing by Huaqing Palace* [*in Chang'an*], *in three verses in the jueju style*), by Du Fu, the Tang poet. They refer to the famous story about fresh lychees, a rare delicacy in the north of China, that were brought from far in the south to the capital at Chang'an (present-day Xi'an) by relays of fast horses to the Imperial Concubine Yang. Known as Yang Guifei (719–756), she was one of China's Four Famous Beauties and favorite consort of the Tang dynasty's Emperor Xuanzong (685–762, r.712–756). Later, while fleeing the An-Shi Rebellion with the imperial entourage, she was accused of extravagance by the emperor's soldiers and the Emperor allowed her to be killed as a scapegoat.—trans.

When I was working in Guangyuan, I proposed and promoted the establishment of e-commerce business platforms for Guangyuan's products. The aim was to use e-marketing to allow Guangyuan's plentiful and excellent edible fungus, red-hearted kiwi fruit,[10] walnuts, vegetables, and other quality agricultural products to have better access to large markets, thus ensuring both high quality and reasonable prices.

At present, many agricultural enterprises have established their own e-commerce platforms or online shops. Some e-commerce businesses have also taken advantage of the Internet to vigorously expand sales of their agricultural products. The strengths gathered by the "wings" of e-commerce have "lifted" the convenience of buying and selling. Agricultural products from the deep mountains that were previously known to only a few can now enter vast new markets. Furthermore, consumers like you and me are able to access more varieties conveniently.

The other story that left a deep impression on me was that of Big Sister Yan, who was a farmer living on Cengjia Mountain in the Chaotian District of Guangyuan. She grew vegetables in the open fields on that high mountain. By organizing a vegetable cooperative that established a brand of vegetables called *Yan dajie* ("Big Sister Yan") she spurred neighboring villages to plant green, high-quality vegetables. They utilize modern logistics to sell these vegetables to the markets of Chengdu, Chongqing, Xi'an, and other large cities. The local farmers have made fortunes from the vegetable produce business due to three factors: cooperatives, brands, and logistics. I have eaten hot peppers and pea shoots from Cengjia Mountain—they really were so fresh, sweet, and tasty. Aside from that, newly emerged investment methods such as crowd funding and management methods such as personal, custom-made farms and other modern innovations will bring about more and more profit for farmers.

4.4.4 Bright Future for Multifunctional Agriculture

As I pick chrysanthemums beneath the eastern fence,
I gaze idly at the southern mountains ...[11]

Farming is an occupation that brings one pleasure, as one works on the land under the sky and sees flowers bloom and fruit ripen. Farming is something that can provide multiple functions: relaxation, touring and sightseeing, culture, and the

[10]The kiwi fruit, a term first used in 1965, is indigenous to China, where it is known as *mihou tao* (literally, "macaque peach"); in English, it has been known since the mid 1920s as the Chinese gooseberry.—trans.

[11]From the poem *Yinjiu* (*Drinking wine*) by Tao Yuanming (c.365–427) of the Eastern Jin dynasty. This is yet another poem that needs no citation for Chinese readers.—trans.

combination of heaven, earth, and man. Urban dwellers seldom see scenes of agricultural production and the natural scenery of rural areas. Touring of farms and rustic villages not only represents an aspect of multifunctional agriculture, but such agro-tourism is also one of its economic growth areas. Everyone is well acquainted with farm home-stays,[12] one of the classic examples of this.

I have two friends who cultivate leased land in the suburbs of Beijing. The plot is not big, being only a few *fen* in area. On weekends and other leisure times, they take their child and the whole family works on the land, experiencing the delight of cultivation and the joy of harvesting. For adults and children alike, it is a form of exercise and relaxation.

In recent years, mainly in suburban areas, some farmers and enterprises that have vision and management ability have devised a brand-new agricultural business mode based on leisure. Utilizing their own or leased land, they level the fields and equip them with irrigation and other infrastructure. Then they divide large parts of the fields into small plots with each parcel ranging from dozens to hundreds of square meters in size. Afterwards, they advertise and promote the leasing of these plots to city dwellers. Taking food safety and leisure into consideration, some urbanites are delighted to spend money to lease land and till it themselves. Both parties thus get what they need.

In Japan, those city dwellers who own and cultivate farmland are much envied. The difference with China is that farmland in Japan in general is owned outright and is not leased. A Japanese researcher at Japan's Comprehensive Research Institute of Agricultural and Forestry possesses some land in a rural area. He feels very proud to return to his village to cultivate the land now and then. No matter whether the land is one's own or is leased, leisure farming, with the help of creativity and marketing, is a new form of agriculture that in the future is sure to promote the merging of urban and rural areas and encourage a return to nature.

Middle Rock Farm in New Zealand is a classic foreign example of leisure-oriented agricultural tourism. Charlotte Rietveld, who is around thirty years old, assists her parents in running a 15,000 *mu* [i.e., approximately 2,500 acres or 1,000 ha] ranch that raises 7,000 Corriedale sheep. As she explains it, because the international wool market is not as good as it used to be, the price of wool has fallen. The ranch operation earns only middling profits and they cannot make as much money as raising milk cows. Like many ladies, Charlotte also enjoys going shopping, but her home is in a remote village, far from a city. Nevertheless, she greatly enjoys her work with livestock and her rural life and feels blessed that she and her family have been endowed with abundant resources and a beautiful environment. She happily rounds up the flocks of sheep with her sheepdog to let them graze, and enthusiastically receives every guest who comes to the farm to take a tour.

[12]Known in China as *nongjiale* (literally, "Happiness of a Farm Home"), farm home-stays are popular with many Chinese city dwellers who want low-cost vacations in rural areas close to nature.—trans.

At any rate, it is interesting and challenging to combine knowledge of farming culture together with agriculturally related science and technology, geography, soil, weather, and history into a business of agricultural sightseeing and tourism. At the same time, to do so, it is necessary to increase overall planning and integration of domestic agricultural sightseeing, rural tourism, and service standards.

Appreciation of Poems about Agriculture

Gui yuan tian ju (qi-yi) (*"Returning to my old homestead amidst the gardens and fields, first stanza)*

—Tao Yuanming (c.365–427), Eastern Jin dynasty official and literatus

Ever since I was a child, I never followed vulgar style; my own inclinations leaned toward an ardent love of nature.
By mischance, I fell into the snare of taking up an official's career; in the blink of an eye, it has already been some thirty years.
A caged bird yearns for its old forest home; a fish trapped in a shallow pond is homesick for its deep pool.
I wish to bring under cultivation wasteland adjacent to my southern fields, to maintain my simple and unadorned nature, and return to farming my fields and garden.
My dwelling place is surrounded by more than ten mu *of farmland; the thatched cottage has eight or nine rooms.*
Elm and willow trees shade the rear eaves; the peach and plum trees vie for attention in the front courtyard.
In the distance, neighboring cottages are faintly visible; in the village, smoke from cooking fires spirals upward.
A dog barks from deep in a small back lane; a cock crows from atop a mulberry tree.
Within the courtyard, there are no worldly affairs to disturb me; in my quiet room, there is only a surfeit of leisure.
After being confined for so long in the chicken coop of officialdom, today I finally return to the forest and mountains.

4.5 Various Supports of Government Policy

4.5.1 Risk Management in Agriculture

The intensification of agricultural risks has caused an increase in uncertainty and serious losses to production and management. There is a need to implement risk management strategies, strengthen risk management systems to reduce harm, and mitigate losses in order to protect modern agricultural development. In terms of management of natural risks, there should be "subtractions," i.e., a lessening or decrease of damage from disasters. Entreating Heaven to bestow a bountiful harvest is an obsolete practice; we must rely on ourselves.

First, we need to fully utilize modern scientific knowledge of meteorology, hydrology, geology, and botanical protection in order to monitor and predict the occurrence of agricultural calamities. Second, we must take precautions by strengthening systems of disaster prevention, preparation, protection, and relief; strengthening systems of preventing animal disease epidemics and crop damage from insect pests; and perfecting emergency reaction mechanisms and long-term cooperative mechanisms of disaster reduction and relief at all levels of government.

In addition, we need to unceasingly improve basic facilities and conditions of agriculture, thereby increasing our ability to resist disasters.

Finally, we also need to step up efforts to support agricultural insurance. In the realms of finance and tax policy, we need to increase the varieties and coverage of agricultural insurance, and alleviate farmers' post-disaster losses.

In terms of market risk management, there should be "additions"; that is to say, governments should add more support and protection. It has already unveiled some adjustment policies in order to stabilize agricultural markets and protect the incomes of farmers working in the fields. The policy of guarantee minimum price supports for grain is intended to protect farmers who would otherwise lose money by selling grain below cost. The grain stockpiling system is intended to solve the problem of farmers having difficulty in selling their crops during years of bumper harvests. Furthermore, the policy of stockpiling pork is intended to stabilize the live hog market. These government policies have all had some positive effect and should be adhered to and refined. Governmental departments of agriculture and commerce at all levels should: establish and strengthen supply and demand mechanisms, price forecasts, risk predictions, and early warnings; provide guidance on production and management of farm products; and not be caught blindsided. At the same time, there should be continuous refinement of the agricultural market system in order to keep product distribution running smoothly and easily throughout the market. There should be effective joining up of farmers with enterprises; producers with sellers; brokers with supermarkets; and areas of production with places of consumption, thus ensuring farm products are sold smoothly and profitably.

Aside from all this, there should be "multiplications" of and "additions" to jobs in agriculture. That is to say, many more workers need to be hired in order to promote the integration of primary, secondary, and tertiary industries that will enlarge the agro-industrial chain and increase-value chain.

4.5.2 Controlling Costs and Stabilizing Prices

To solve the problem of high costs and low earnings in agricultural production, there ought to be a strategy of cost control and appropriate adjustments, constraints on agricultural costs to prevent them from rising too quickly, and a rational increase in the price of farm commodities. The following measures should be taken:

First, control the upward momentum in the price of petroleum and other primary products, increase the efficiency of their production and distribution, and make great efforts to ease the price increases of farming materials that result from the price increases of primary products.

Second, support the production and distribution of farming materials along the production chain (from the aspect of public finances, tax revenues, loans, and management), and lower the cost of those materials.

Third, expedite development of various advanced, suitable agricultural machinery to substitute for the continuing decrease in the size of the agricultural labor force, in order to cut down labor costs.

Fourth, actively carry out construction of irrigation and water conservancy infrastructure and land reclamation work, and improve land and labor productivity.

Finally, strengthen the dynamic equilibrium mechanisms that affect agricultural costs and the price of agricultural products. In the face of rising costs, maintain a reasonable increase in the prices of farm products. Maintain reasonable prices for agricultural products and establish subsidies for agricultural products that are both scientifically reasonable and operationally convenient.

4.5.3 Increasing Agricultural Investment and Subsidies

Even if all the land in the world were under cultivation,
Farmers would still starve to death.[13]

Agriculture is a feeble industry, and farmers are the underprivileged.

Increasing investments and subsidies are the main ways to support agricultural development and increase agricultural profits. On January 1, 2006, China officially abolished the *Agricultural Tax Regulations* that had been passed in 1958. The agricultural tax system thenceforth ended. Although due to limiting conditions, this happened later than in many developed countries, but to farmers the good news finally arrived. According to statistics, after the complete abolition of agricultural taxes in 2006, the tax burden that has been alleviated each year is 125 billion RMB (using the year 1999 as a baseline for comparison), which is almost equivalent to the expenditures spent on agriculture by the Ministry of Finance at that time. Through agricultural tax reform, state finance has gradually been increasing the amount spent on agricultural investment. Farm subsidies are appearing one after another. China's agricultural tax system has achieved great transformation from a system that took, to a system that gives, and from one that collected agricultural

[13]This is part of a couplet from the poem *Minnong* (*Pity the Peasant*), by the Tang dynasty literatus, Li Shen (772–846). For Chinese readers, it needs no citation.—trans.

taxes, to one that grants agricultural subsidies and increases agricultural investments.

In 2004, as soon as the Central Committee of the CPC and the State Council decided to lower and then actually abolish agricultural taxes, for the first time the Ministry of Finance withdrew money directly from the grain risk fund to provide direct subsidies to grain farmers in thirteen major grain-producing regions. Furthermore, the farmers in certain regions received subsidies for purchasing improved varieties of seeds. Finally, some received subsidies for purchasing agricultural tools and machinery. During that time, these three direct agricultural subsidies amounted to 14.5 billion RMB.

In 2006, to solve the problem of the increasing cost of agricultural materials, the nation increased comprehensive subsidies on agricultural materials. Henceforth, there have been mainly four agricultural subsidies: direct grain subsidies; subsidies for the purchase of agricultural equipment; subsidies for farming materials; and subsidies for improved varieties of seed or livestock. China's agricultural subsidy system has gradually been improved and has seen yearly growth. Overall, farming is now gradually entering an era of "High subsidies."

In 2013, these "Four Agricultural Subsidies"[14] supported by the Ministry of Finance exceeded 170 billion RMB, more than a ten-fold increase over 2004. In addition, the government has considerably increased funds that are used to reward counties that produce large amounts of crops. Meanwhile, in major grain producting areas, it canceled completely the requirement that local areas provide matching funds for the government risk funds. It expanded local grain risk funds. It arranged for subsidies to pay premiums on agricultural insurance. It proposed a series of projects to provide subsidies for sows that are used for breeding and milk cows. In addition, it provided for improved varieties of livestock and poultry, prevention of animal epidemics, and soil testing for determining the correct application rate for chemical fertilizers.

Moreover, one must mention that to repair, protect, and manage the ecological environment, in 1999 subsidies were launched to encourage farmers to allow some agricultural land to revert to forests and in 2003 to reconvert some pastureland to wild grassland. These comprehensive and specific government subsidy policies have had important effects on supporting agricultural production and motivating farmers. In recent years, state investments in agriculture have steadily gone up, demonstrating "High investment."

In 2012, government expenditures on the "Three Rural Issues" [viz, agriculture, rural areas, and farmers] was 1.23876 trillion RMB, an increase of ten times more than was spent in 2000, and which accounted for 9.8% of government expenditures. This last figure was 2% higher than that in 2000. Of this, 478.51 billion RMB went to support agricultural production, 164.3 billion RMB was spent to support the "Four Agricultural Subsidies," and 533.91 billion RMB was spent on rural social development. This was not merely a considerable increase in terms of absolute

[14]Mentioned in Chap. 1, Section IV, Subsection 2.—trans.

value, but also represented a steady increase in the proportion of government expenditures.

Although in recent years there have been "High investments" and "High subsidies." the problem of "Low effectiveness" still exists. Moreover, because financing agricultural infrastructure is deeply in debt, irrigation and water conservancy facilities are still weak, and two-thirds of the nation's farmland probably still only provides low or medium yields. Hence, it is difficult to support further increases in the output of grain and other agricultural products. Subsidy incentives for grain farmers in large grain-producing areas have gradually increased, but they are still lagging behind, whether it is compared to developed countries that enjoy high subsidies or is compared to practical needs, such as grain production and farmers' income.

Therefore, it is necessary to: steadily increase investments and subsidies in the future; implement strategies for publicly financed investment and supplies; optimize policies on investment and subsidies; and increase the efficiency of subsidies. Central and local finance should further increase the absolute value and proportion of financial expenditures on agricultural investment and policies should be established that encourage profits from the sale of land to be invested in rural land development and basic facilities and construction.

Furthermore, the county should do the following: expand agricultural investment channels; encourage social capital to invest in agricultural infrastructure; and comprehensively develop land regulations. Continue to strengthen financial support to main grain-producing regions; increase financial and grain production subsidies to these granary regions; and increase capital incentives to major grain-producing counties. Furthermore, improve agricultural subsidy systems and policies. Steadily increase the "Four Agricultural Subsidies" and the funds to pay for them. For the main food crops, implement preferential subsidies for improved varieties of seeds. Expand the scale and variety of subsidies for buying agricultural implements. Raise subsidies for rural farming materials. Finally, establish a dynamic adjustment mechanism for comprehensive agricultural material subsidies that is directly coupled to the rise in prices of farm materials. In summation, the adoption of high investment and high subsidies will promote the smooth drive toward modernization of Chinese agriculture.

From our research in a number of locales, we found that invariably there are cadres at the grassroots level who report that farm subsidies are like "sprinkling in peppers" when cooking [i.e., subsidies are an unduly inefficient distribution of limited financial and material resources, without regard for priorities]. Their view is that scattering subsidies on every plot of land will not result in putting any real money into the hands of the farmers; hence, it would be better to combine the many small subsidies into fewer but larger and more meaningful ones. Indeed, due to the relatively small scale of Chinese farm households, which constitute a large base of over 200 million, it is true that individual farmers directly receive very little in agricultural subsidies.

But I would rationalize this policy in this way: First, subsidies directly benefit farmers who are engaged in production. At the very least the governmental framework of farm subsidies has been established. Developed countries in the West have also done this. Afterwards we can continue to refine the subsidies (for example, we can preferentially provide subsidies to households that operate on a large scale). Second, although up to now farm subsidies have not really been that much, they have increased yearly and will continue to increase. Third, let other things [e.g., various agricultural infrastructure projects, etc.] be funded through other channels.

Finally, I would like to tell a story that I heard from Director Han Jun. Several years ago, someone wrote a report for a high-level official on the problem of subsidies. After listening to the arguments, the official wisely responded, "In giving benefits to farmers, you have given them. So how exactly do you propose taking them back?"

Chapter 5
Who Will Till the Land and How Will They Till It?

You plow the land and I'll weave the cloth. I'll carry water on my shoulder pole and you use it to irrigate the fields.
—Tianxian pei (Marriage of a Fairy Princess) (This is an important operatic folk legend from the repertoire of Huangmei Opera that originated in Anhui Province.—trans.)

Recalling the other lyrics of this song depicting the production scene of men plowing and women weaving, as well as the descriptive phrase, "thirty *mu* of land and a head of oxen,"[1] cause me to reflect on small farmers and small farms of the past. It has been more than sixty years since China's liberation [i.e., the establishment of the People's Republic of China in 1949] and more than thirty years since the rural reforms. What is the present state of small farm agriculture? What problems exist? Are there any new phenomena related to these?

"Small farms are in a predicament" is how I would summarize the dilemma of the low level of development of small-scale agriculture and small-scale farming household operations. The term "old people's agriculture" has gradually become more prominent in recent years. It describes the phenomenon of large numbers of young and middle-aged laborers who have left their farms, leaving behind mostly elderly farmers to manage things. Farm household operations are small-scale and those who till the land are old—"small" and "old" are the two operative words that bear on much of today's Chinese farming and its management methods. This leads to the question of, "Who will till the land and how will they till it?"

[1]This is part of the well-known phrase, "Thirty *mu* of land and a head of ox, a wife, children, and a warm bed," from a 1947 speech by Liu Shaoqi (1878–1969), an important Communist Party leader. –trans.

Y. Zhang, *Insights into Chinese Agriculture*,
https://doi.org/10.1007/978-981-13-1050-8_5

5.1 Who Is Cultivating the Land Now?—Old Farmers

5.1.1 Concept of "Farmer" Needs to Be Sorted Out

The literal meaning of the word "farmer" is "a person who is engaged in agriculture and who does farming." However, this word carries other meanings as well. China has implemented a dual urban-rural permanent household registration system under which people born in cities are classified as "city residents" and those born in rural areas are "farmers." Therefore, the term "farmer" also carries the meaning of one who is born in a rural region and thus it is a kind of birth identity or social status. Consequently, the term "farmer" differs in what it refers to; sometimes it refers to what one does and sometimes it refers to one's origin.

I will give you an example. When I was in high school, my mother often instructed me, "Study hard and strive to pass the university entrance examination. Then you won't have to work as a farmer, browbeating the clods of earth in order to make a living." The word "farmer" in this sentence refers to an occupation; by going to college, one's household identity can be changed from "farmer" to "city resident." Furthermore, one could choose to do non-farm work and consequently enjoy a higher level of public services and social welfare.

In July, 2014, the State Council issued *Suggestions Regarding Further Promotion of Reforms of the Household Registration System*, which combined the urban and the rural household registration systems[2] of cities and rural areas, implementing a unified resident permit system. It was expected that the reform of this, the *hukou* system, together with other institutional reforms, would transform the duel systems of urban and rural areas [This would thus ameliorate problems faced by farmers who move to cities only to find that the public services and social benefits that are available to them are unequal to those of urban residents who were born and grew up there.]

The term *nongmin gong*[3] refers to those who left their farms to perform manual work and whose household registration still states that they are farmers. In the wake of economic development and the opening up of labor markets, more and more laborers with "farmer" as their identity have been "de-agriculturalized." The majority of them have left the farm to do manual labor, living and working in cities and towns. So, what is their identity? Are they still farmers? The answer is yes and no. It ambiguousness that has given rise to the neologism *nongmin gong*, that specialized term referring to manual workers in cities and towns who are registered as farmers.

In 2013, Chinese migrant workers numbered some 269 million; of these, 166 million were working outside their home villages. The majority of them have never

---

[2]This term is often referred to in English as the *hukou* system (or sometimes as the *juji* system), China's system of registered permanent residence.—trans.

[3]*Nongmin gong*, literally, "farm laborers," is usually translated in English as "migrant workers" to differentiate them from farm laborers who remain on the farm to work.

undergone any occupational training. In the cities, they mainly enter manufacturing, construction, or service industries. "Migrant worker" is something of a misnomer: from their household registration identity, they appear to be farmers; but from what they do, they are workers. By analogy, should there not also be the concept of *nongmin shang* ("farm merchant"), which would mean a former farmer who has now gone into a non-farming business? [Of course, there is no such term.]

Who is a real farmer?

By the end of 2012, the population of Mainland China was 1.35 billion. According to the statistics of the Ministry of Public Security, 65% of the population was registered as agricultural residents; those registered as farmers numbered approximately 880 million. According to the National Bureau of Statistics, among all agricultural residents, the number of people living year-round in rural areas (defined as those registered for six or more months per year) was 640 million people, accounting for 47% of the total population. This means that the population of rural residents was smaller than the rural-registered population by 240 million. Of the 640 million farmers in rural areas, the number of laborers was close to 400 million people; some 260 million of them were engaged in primary industry [i.e., farming], including both professional and part-time farmers. Moreover, part-time farmers outnumbered professional farmers. Over the last thirty years, there has been a downward trend in the absolute number of China's agricultural workers and in their ratio to the total number of workers. The trend in the decline in the ratio of agricultural workers to other workers has been especially pronounced, falling from 68% in 1981 to 37% in 2010.

5.1.1.1 Knowledge Link

At present, the National Bureau of Statistics calculates the rural and urban populations based on resident population. "Resident" is defined as one who has resided in one place for one year or for a total of six months or longer within that year. This statistical approach began in 2000 with the fifth nationwide population census. Previously, rural and urban populations were calculated on the basis of the household-registered population [viz, the hukou system]. Everyone knows that the percentage of urbanization equals the urban population divided by the total population. After the change in statistical approaches in 2000, it became much easier to calculate this percentage. Previously, however, it could only be estimated by using some complicated methods. Written before 2000, my MA thesis on urbanization took me a lot of work to figure out the percentage of urbanization.

5.1.2 Old Farmers Are Busy in the Fields

If you go to the countryside, you will discover scenes of "Old people's agriculture"; most of those working the land are middle-aged or elderly. According to statistics, more than 40% of China's agricultural laborers are over fifty year old. Those over sixty years old who are still doing farm work, and even those over seventy, can be found everywhere. A report by Cao Liqun and other researchers at the Policy Research Office of the CPC Central Committee mentions that according to the standards set by the related organizations in the United Nations, those workers aged forty-five years and above are counted as older workers. When the proportion of middle-aged and older workers accounts for 15% or more of those in the total labor force, then the labor force is considered be aging.

China's agricultural labor force is already clearly aging. It is predicted that by 2016 more than 50% of China's agricultural workforce will be fifty years or older. But according to government standards, men sixty years old and above and women fifty-five years old and above should not be counted as part of the workforce. More than twenty years ago, most of the people cultivating crops, from plowing to reaping, were farmers in the prime of their lives. But today, older farmers have already become the main force of farm production. In urban enterprises and institutions, men at age sixty and women at age fifty-five reach retirement age. Old farmers have no such leisure; they still work in the fields to plant crops.

China has about 125 million new-generation migrant workers [born since 1980]. Basically, they cannot farm and in the future they will not make a living by farming. The aging of its labor force is becoming a major problem facing the current and future development of agriculture.

As a researcher, Han Jun told this story. In 2006, a scholar from the UN Secretary-General's office was conducting an inspection in the Ningxia Hui Autonomous Region. He observed that the profile of its rural population was exactly like that of a country during wartime, because there were no young men in sight. There were only old man, women, and children; the young men had all migrated elsewhere to work. This can now be seen even more clearly in the mid-western region of the country. Furthermore, in the developed Eastern region of the country in the city of Suzhou, more than 90% of the farm labor force has achieved stable employment in non-agricultural industries. Not just farmers in the prime of life are engaged in non-agricultural industries, but there are also some middle-aged and elderly farmers doing so.

Appreciation of Poems about Agriculture

Yelao ge (Song of the old peasant)

—by Zhang Ji (767–830), Tang dynasty poet

An old peasant lived in an impoverished house deep in the mountains;

He had managed to cultivate three or four mu *of mountainous fields.*

But sprouts were few and taxes many, so there was nothing left for his family to eat;

His tax grain ended up wasted in the government granaries, where it rotted into dirt.

At the close of the farming season, the plow was leaned against the outside of his humble abode.

He called to his son to climb the mountain to gather acorns.[4]

The rich merchants of Xijiang[5] *have countless pearls and jewels;*[6]

On their barges, even the dogs they raise can frequently eat meat.

This poem depicts an old farmer in a mountainous district during the Tang dynasty. Under heavy taxation and with little land, he and his family are forced to eat acorns to allay their hunger. It also shows the large socio-economic gap between rich and poor.

5.1.3 Young and Middle-Aged Farmers Have Rushed into Cities

The data taken from the second national survey of agriculture shows that in Luzhou, a prefecture-level city in Sichuan Province, as early as 2006, ninety-one percent of the rural labor force under the age of forty had left their home villages to work somewhere else. A questionnaire survey in 2013 conducted by the provincial department of agriculture indicates that a mere 15.8% of those below forty years of age were engaged in farming.

[4]Acorns were a survival food during times of hunger.

[5]Xijiang was the area around present-day Jiujiang in Jiangxi Province.—trans.

[6]Literally, "a hundred *hu* of pearls," with a *hu* being an ancient dry measurement of volume, also called a "peck."—trans.

Some readers may wonder, where have these younger farmers gone and what have they been doing? Of course they have migrated to other places to make money by doing manual work or business! At the moment, these are the main choices for young farmers. Why is this? It is because agricultural benefits are relatively low. The income from a year's work in the fields is less than what can be earned in a single month of hiring oneself out; growing crops makes far, far less money than working for others. People will always seek to improve their lives and pursue happiness; it is normal for those people to migrate to areas and occupations where they can earn higher income. That is why young farmers have migrated one after another to make a better living.

In recent years, it is even clearer that urbanization and the development of industry and the service sector have had the effect of draining more and more young farmers from rural areas. This clearly explains the reason why there is an annual "shortage of migrant laborers" at the start of the Chinese New Year [as they return in their tens of millions to their home villages to pass the two-week Spring Festival with their families.] The security guards in Beijing are no longer young people, as they were a few years ago. For example, the two guards in the residential community where I live are both more than sixty years old, coming from farming villages in Henan Province.

5.1.4 Difficult to Modernize "Old People's Agriculture"

"Old people's agriculture" is difficult to modernize. Nowadays most of the farmers who are actually cultivating the land were born before the 1960s. In general, their educational level is low and thus the quality of the labor force is below average. In 2010, data from the sixth nationwide census shows that 93.6% of those working in farming, forestry, animal husbandry, and fisheries never went to middle school. They have a weak ability to embrace modern agricultural technology and information; in most cases, they are forced to passively accept advanced technology, information about the sale of farm commodities, and knowledge about business management and administration. Yet, modern agriculture requires not merely workers who are engaged in production, but also those who can understand both farming and business operations and who are adept at utilizing modern farming methods and information technology. An aging labor force impedes the adoption of modern technology and the use of advanced materials, facilities, and equipment. It thus constrains the development of modern farming operations.

This is not to look down on elderly farmers. My grandfather is almost ninety years old this year. Ten years ago, he was still growing crops; moreover, he is a competent man who is adept at the intensive and meticulous cultivation that is typical of traditional farming. However, it is true that he has not stayed abreast of agricultural modernization.

It is certainly difficult to accomplish modernization of agriculture while depending on a labor force that is graying and has little educational. "Who will till the land?" and "How will they till it?" have already become realistic questions. The answer to these questions is that we must provide vocational training for farmers as soon as possible. No matter how good agricultural development plans and government policies are, how advanced agricultural production technology is, or how well equipped farmers are, in the end everything depends on farmers to embrace and use mordern technology. Only professional farmers can shoulder the important responsibility of developing up-to-date agriculture.

5.2 In the Future Who Will Till the Land?—Professional Farmers

"Professional farmer" refers to one who takes agriculture as an occupation. It has the identical meaning as the word "farmer" with the added word "professional" to distinguish it from the concept of someone who is listed as a "farmer" just because he is registered that way under the household registration system (or whose permanent address is in a rural area).

However, under the current situation where we must modernize agricultural even as large numbers of the rural labor force have migrated to perform manual labor or to do business, the term "professional farmer" should be given new meaning. This can be summarized as one who is: knowledgeable about technology, competent in business, and has become professionalized. Knowledgeable about technology means mastering farming technology, including both traditional and modern modes. Competent in business means being able to manage marketing as well as cultivating the fields. Professional means taking agriculture as one's sole or main occupation, rather than manual labor or business.

5.2.1 Modern Agriculture Must Depend on Professional Farmers to Till the Land

Of all the factors involved in production and development, people are the most basic. Agricultural modernization requires professional farmers to play the major role.

Take American agriculture as an example. Professional farmers manage it all. Farmers there make up only about 1% of its more than 300 million population; there are approximately 3 million farmers in total. American farmers may be truly called professional farmers, with a high educational background, technical level, and degree of specialization. They are able to operate modern implements and equipment, can practice scientific breeding, and are experienced in using the Internet to obtain market information. These are important reasons why only 1% of America's population supports not only their own country, but makes America the largest agricultural exporter in the world. Furthermore, the income of American farmers is quite decent: according to reports from America's Department of Agriculture, in 2013 American farmers' net income reached $128.2 billion, with per capita income of roughly $40,000 or about 240,000 RMB.

Professional farmers also manage Japanese agriculture. Its agricultural workers make up only 5% of its total workforce. The number of farm households has dropped from 6.057 million in 1960 to only 2.528 million by 2010. Among them, 1.632 million were farm households that also marketed their products (this refers to farm households that managed cultivated areas of 450 *mu* or above or had annual sales volume of at least ¥500,000 [about $4,500]); the rest were small-scale, self-supporting farm households. Among those market-oriented farm households, professional ones totaled 452,000, with the rest being part-time.

Japan also faces the problem of "old people's agriculture." The aging of its agricultural labor force is critical; in 2012, the average age of farmers was sixty-six years old, 60% of whom were age sixty-five years or older, and those under forty made up mere 10%.

For a long time, the main factors holding back agricultural modernity in our country have been the low quality of its labor force, the poor development of specialized vocational education, and a lack of training for farmers. In 2012, only 15.6% of China's farmers possessed a diploma from a high school (i.e., a vocational secondary school) or above. The rest of the labor force was educated to the level of junior middle school or less. Approximately 70% of China's agricultural labor force has not received any technical training; a great number have never received any vocational education or training and therefore lack skills.

Cultivating professional farmers and boosting modern agricultural development requires strengthening vocational education and training. Farmers are the ones who put into practice development and who promote modernization. Promoting new technology, increasing the degree of mechanization, and enhancing specialization—all place higher demands on farmers. Farmers can no longer simply rely on traditional farming methods. Rather, they must update their knowledge and technical skills to promote the transformation from traditional farming to modern agriculture. At the same time, the rural economy and social structure have already undergone tremendous transformations. An expanding economy and an increase in production have placed new demands on farmers. These all require relying on farmers who have received vocational education and training.

5.2.2 How to Foster Professional Farmers?

It is necessary to formulate and put into effect a strategy to foster professional farmers. There are two ways of doing this. One is by vocational education, which refers to a relatively systematic and regular school education. The other is vocational training, which refers to special training offered by governmental departments, social organizations, and enterprises. In recent years, our country has already promulgated a series of policies that support farmers' vocational education and training. Professional education and training have gradually been emphasized and strengthened, and have made great headway. The scale of professional education has expanded unceasingly and the form and content of training has been enriched. However, there have been some problems regarding vocational education and training. Therefore, these two aspects need to be further strengthened.

First, we must formulate a national strategy cultivating professional farmers and lay heavy emphasis on this. To do so, we must make comprehensive plans, carry out strategies, propose strategic tasks, map out step-by-step objectives, and take concrete action. With regard to the broad countryside, professional farmer training serves the needs of local agriculture. Local government agricultural departments ought to be more farsighted and realize the significance of vocational agricultural education and training, regard it as their primary task, and give full support to aspects such as policy, funding, and school management.

We should reinforce education and training of professional farmers; provide free training to current farmers through ways such as disseminating agricultural technology via TV broadcasts to schools; and emphasize training of those major breeders and farmers who have managerial abilities.

We should provide potential agricultural workers, especially those rural middle or high school graduates who intend to go into the profession of agriculture, with free vocational education and training. We should also increase support for professional farmers and provide comprehensive training in the areas of financial subsidies, loans, insurance, investment, technical services, and social security, allowing professional farmers to obtain the same or even higher salaries than those workers engaged in non-agricultural industries. This would arouse enthusiasm for becoming a professional farmer or for getting involved in farming and management.

Second, we should collect funds from multiple sources and establish a firm financial basis for the strategy of fostering professional farmers. Professional education and training for farmers is the same as general education, in that they both are public and depend on external support. Government finance should actively support vocational education and training, especially considering that in the past there was seriously insufficient investment in this area. In comparison to general education, vocational education requires even higher investments.

Furthermore, government should adopt incentive policies and set up incentive mechanisms to actively guide enterprises and social capital to invest in education and training for farmers. On the one hand, we should encourage social capital to establish vocational schools and training institutions in rural areas; support enterprises and individuals that participate in such vocational education and training; and provide appropriate preferential measures in obtaining authorization and required official endorsements, taxation, financial subsidies, credit, etc. On the other hand, we should establish a mechanism that involves multiple parties (i.e., government departments, social capital, business, and farmers) to actively invest and participate in farmers' professional education and training. This could take the form of governmental departments, social capital, business, and farmers all actively investing and participating in diversified investment mechanisms for farmers' professional education and training.

Third, we should fully utilize the resources of different government sectors, such as agriculture, educational, labor, science and technology, and poverty alleviation, and focus this united force to carry out vocational training strategies for farmers. Comprehensively applying resources of government departments and coordinating social resources are an important aspect of this task. There should be structures involving management and implementation of multiple departments (e.g., agriculture, education, labor, science and technology, and poverty alleviation) to share the limited amount of government resources.

In the process of promoting vocational training for farmers, there is an issue worth paying attention to. That is, how to break through departmental limitations in order to utilize fully their resources, avoid duplication and waste, and enhance the efficiency of the use of financial capital. The central and local government departments urgently need to innovatively set up systems; comprehensively utilize all categories in rural education, training, and science and technology; and promote agricultural technology. At the same time, there needs to be coordinated use of government and social resources and full use of market forces to allocate resources. The needs of the market and farmers should be respected, and there should be no resort to coercion or commandism. The departments in charge of government oversight must do the following: Encourage social resources to participate actively in professional education and training. Keep management separate from actual educational training. Initiate a system of market competition. Optimize allocations of training resources. Fully mobilize different departments, different industries, and different types of government and social resources to participate in rural professional education and training. Finally, they must form a united force to carry out the fostering of professional farmers.

5.3 How Has Land Been Tilled Since Ancient Times?—By Small Farm Agriculture

5.3.1 What Is Small Farm Agriculture?

"Small farmer" is a concept of agricultural economics. Research into "small farm agriculture" and "small farm economy" is an important part of agricultural economics. However, these three concepts are hardly a part of the government policy lexicon, and are instead replaced with "small-scale farm household management" or "small-scale agriculture," etc. I personally believe that from the angle of research and narrative, a starting point is to use the terms "small farm agriculture" and its variants, because they are more down-to-earth and accurate. Using these terms is also easier when discussing the main aspects of farm management.

The Chinese term *xiaonong* means "small farm household" (or "small farmer"); this is the simplest and most literal meaning of the word. "Small" means that the operational scale of the farm is small. Its economic significance is small as well; no other meanings are implied. Nevertheless, small and large are relative; a small farm and a large farm (the latter of which is also called a *dahu*, "major producer") are also relative. Does tilling a hundred *mu* constitute a small farm or a large farm? In Heilongjiang Province, it would be considered an ordinary small farm, while in other areas of China it would be considered a large farm; but if it were in America, it would be considered a tiny farm.

Therefore, what is a "small farmer?" He is an individual farmer who relies on his own labor to engage in small-scale farm production. This is the definition found in the 1982 edition of the *Jingji dacidian, nongye jingji juan* (*Dictionary of economics, volume on agricultural economics*). In the *Collected Works of Marx and Engels*, "small farmer" refers to those who are owners or tenants of small plots of land—in particular, it usually refers to owners whose land is generally not large enough to absorb the efforts of the whole family to cultivate it, but is large enough to provide for the family's essential needs. To paraphrase the definition of "small farmer" found in *The New Palgrave Dictionary of Economics*: A small farmer is someone who cultivates the land. Whether or not he possesses absolute rights to his own small plot, he has land to plow.

The definitions of "small farmer" in a few other economic dictionaries also focus on the following two special features. The first is the small-scale of agricultural production. How small is small? I believe the criterion put forth by Marx and Engels is worth adopting: small is considered any scale smaller than what a total farm household can cultivate and manage. This is a very important concept for defining "small farmer."

The second feature is that farm production and management is entirely or is mainly done by the farm household members themselves. To summarize the above definitions and analysis, I conclude that the term "small farm" refers to the individual farm household that completely or mainly relies on the labor of its own members to engage in production, the scale of which is within its capabilities. This definition is extremely important to the discussion in this book because if this concept is not clear then there is no way to discuss the differences between small farmers and family farms or the boundaries between small-scale production and large-scale production. Nor can one talk about "the predicament of the small farmer" and the way out of that predicament.

Related to small farms are the terms "small farm economy" and "small farm agriculture." I believe that these latter two concepts are more or less the same and can be used interchangeably. Small farm agriculture refers to small-scale farming that is managed by small farmers. In the definition of "small farm economics" found in the *Jingji dacidian, nongye jingjijuan* (*Dictionary of economics, volume on agricultural economics*), it is also known as *nongmin getijingji* ("farmers' individual economics"). In general, this refers to an economy that is based on family as the economic unit and that completely or mainly relies on the family's own labor to manage small-scale agriculture independently in order to satisfy its own needs of consumption.

My father and forebears were all small farmers; the fact is, I am also counted as a small farmer because I use my own labor to plow and weed a research plot. My

Weibo [a Chinese microblog application] and WeChat accounts both use the name "Glory and Prosperity to Small Farmers" (*Xiaonong ronghua* 小农荣华). It is my heartfelt wish that small farmers, in their vast numbers, all lead happy lives.

5.3.2 *Small Farm Agriculture Is Long Standing and Well Established*

In the later stages of primitive society, individual families started separating from their clans; thus, the farm household became the unit of agricultural production and management and separated from the collective operation under which land was publicly owned by the clan. Henceforth, farm households became the principal farm managers.

So where should we start discussing the small farmer and small farm agriculture? We need to start from examining the scale of household management and the household's ability to cultivate the land.

The textual research in *Zhongguo tudi zhidu shi* (*History of Chinese land tenure systems*), co-written by Chinese American economists Zhao Gang and Chen Zhongyi, shows the approximate average area of land managed by farm households in every Chinese dynasty. Under the well-field system of the Zhou dynasty, public land was allocated to farmers, with a family of five being able to obtain a share equal to 27.3 modern *mu* (of actual tillable land). The unit of land measurement at that time was the "100 pace *mu*," defined as one hundred paces [i.e., a *bu*, or "a double step"] long and one *pace* wide. During the Spring and Autumn and the Warring States periods, the technology of the iron plowshare emerged, and farmers were then able to cultivate more land. The Han dynasty practiced a system of limited private ownership of land, with each farm household managing a land area of sixty-nine modern *mu*. Because of the fallow system in use at that time, the actual sown area was about fifty modern *mu*. This was the upper limit for a farm family to manage. The Western Jin dynasty instituted the *zhantian* system, with each man and each woman on average sowing and paying taxes on fifty-two modern *mu*. The Northern Wei dynasty practiced the equal-field system, with each family sowing the equivalent of sixty-seven modern *mu*.

Zhao Gang and Chen Zhongyi importantly concluded that, from the Qin and Han dynasties until the Southern and Northern dynasties, there were no strict limits on the amount of land that could be farmed. Therefore, the areas that farmers managed matched their work capability. That is to say, during that period of history, the scale of average household management was at a reasonable level, being from fifty to sixty modern *mu*. Due to the limitations of the technology of that time, farmers were not able to handle more land. The scale of agricultural management matched the ability of households. Thus, according to my definition above of small farmers, members of the average farm household of that time would still not be counted as small farmers.

Another judgment of Zhao Gang and Chen Zhongyi is that after the Sui and Tang dynasties the increase in the Chinese population surpassed the increase in arable land; moreover, due to the improvement of farming techniques, including agricultural tools and application of fertilizer, etc., the relationship between man and the land started to become tense. The land gradually became the factor that restricted production; the limits of manpower that previously had restricted it were replaced by the limits of land. It is safe to say that after the Sui and Tang dynasties, Chinese farm households could be called small farmers. From that time on, small farm agriculture and a small-scale peasant economy started to take shape. The Tang dynasty adopted the equal-field system, with each household being restricted to an upper limit of sixty-five modern *mu*, whether fallow or not. Basically, this upper limit of arable land could not be reached [because of a lack of sufficient land]. During the beginning of the Northern Song period, the average land area for a household was forty modern *mu*. Afterwards, following the increase in population, arable land per capita rapidly decreased. The overwhelming majority of the tenant farmers of the Southern Song dynasty cultivated less than twenty-five modern *mu*, and a great number of them cultivated less than eight *mu*. During the Ming and Qing periods, the average arable land per household gradually decreased still further. By 1936, it was only 18.4 modern *mu*, with an average per capita of only 3.6 modern *mu*.

After the establishment of the People's Republic of China in 1949, as the population increased and farming technology progressed, the human-to-land ratio tightened even further. The scale of small farm management became even smaller; at present the average arable land per farm household in China is a mere 7.5 *mu* and is a typical example of small farm agriculture with its lower than reasonable human-to-land ratio. It is equivalent to 1/40 of that in the European Union, 1/400 of that in the United States, and less than half of that in Japan and South Korea. To make matters worse, land in China is chopped up into bits and pieces, with seven to eight plots for each household. On average, a plot of land is only about one *mu*. In places like Guangxi Province, there is a particularly high degree of fragmentation, with many farm households having ten or twenty tiny plots, most of which are only several *fen* in size.

Therefore, my viewpoint is that for over a thousand years through the Tang, Song, Yuan, Ming, and Qing dynasties until today, economically speaking the scale of farm household production has gradually decreased to a low level that is less and less rational. The land that farm households are capable of cultivating is much larger than the land that they actually cultivate. Thus, the surplus of agricultural laborers has gradually increased, while the average number of hours each farmer works has gradually decreased. As a result, farm households have already become typical small farmers. Small farm agriculture has become a major form of operation for farm production, whether the land is owned or leased.

According to Zhao Gang and Chen Zhongyi, in terms of land possession, even during the Ming and Qing periods there were very few landlords with large holdings of land (land owned by the imperial family or government was not included in their calculations). However, there were a number of exceptions or special cases[7] that researchers have pointed out in the past. The following are four pertinent quotes from their book:

1. *Farmland was regularly distributed into the hands of middle and small-size proprietors; while the land they owned was not a lot, after it was distributed and rented out, it provided land to a certain number of tenant farmers. One of the main reasons for the rare emergence of large landlords was that there were many people and not enough land. Accumulating capital in rural areas was not easy and so buying land was difficult.*
2. *It usually took generations, many decades, or even one or two hundred years, for a small or medium-size landlord family to accumulate two or three hundred mu of farmland. Peasants attached great importance to land that they owned; they did not lightly abandon or sell their land. By the 20th Century, large landholders were few and far between. Those so-called landlords were mostly middle or smaller-sized land owners; strictly speaking, they ought to be called "owner-peasants," since they plowed some of their own fields, with the remainder being leased out to tenant farmers.*
3. *The Land Commission (in 1937) surveyed the entire country and found that landlords on average only owned* 34 mu. *Of this* 34 mu, *not all of it was rented out to tenant farmers; on average, landlords cultivated* 7 mu *and leased out* 27 mu.
4. *Because of high pressure from the large population, it was difficult for big landlords to emerge. The majority of landowners were middle and small proprietors. Under conditions of little land and many people, the great majority of peasants' lives were poverty-stricken. Not only were the tenant farmers in straitened circumstances, but so were small landlords.*

Research from the scholar Philip C. C. Huang roughly corroborates this. He found that on the North China Plain in the 1930s, household-managed farmland made up 90% of the total arable land in that area.

Appreciation of Poems about Agriculture

Qingping yue, Cunju (Life in the countryside, set to a Qingping yue *melody)*[8]

—Xin Qiji (1140–1207), Southern Song dynasty poet

There is a grass hut with low eaves beside a flowing creek with verdant grass all around.

[7]Some exceptions to this general rule were the following large landowners: (1) the kings of vassal states during the Ming dynasty, who were enfeoffed with large amounts of land; (2) during the reign of the Qing dynasty's Kangxi Emperor (1662–1722), the Imperial Household Attendant Gao Shiqi (1645–1704) and the Minister of Justice Xu Qianxue, (1631–1694; 3) during the reign of the Yongzheng Emperor (r. 1723–1735), the Governor-general of Zhili Province, Li Wei (1686–1738; and 4) during the early Jiaqing reign period (1796–1820), Heshen (1746–1799) [a notoriously corrupt official], whose property was confiscated.

[8]While the melody itself has been lost, the literal meaning of *qingping yue* is "clear and peaceful melody." It was one of the melodic schemata of *qu* and *ci* styles of poetry that was popular starting from the Tang dynasty.—trans.

A beautiful, old, white-haired couple is immersed in playful, happy banter in their Wu topolect.[9]
Their oldest son hoes beans on the east side of the creek; their middle son just now is weaving a chicken coop.
Their youngest son that they love the most is a rascal; he is lying down resting by the stream, shelling lotus pods and eating their seeds.

From the type of farm operation depicted, this poem shows a joyous and harmonious scene of a peasant family.

5.3.3 Small Farm Agriculture Conforms to National Conditions

Rulers throughout history, in view of the problems caused by annexing land, put into effect policies to restrain it; this was the political reason why small farmers dominated agricultural operations. A related discussion may be found in the book (of which I was the lead author), *Zhongguo nongdi liuzhuan wenti diaocha* (*Investigation into farmland transaction issues in China*), in the section *Jianbing yu yizhijianbing*: *tudi liuzhuan de san-qian nian lishi* (*Annexation and restraints on it*: *3,000 years of land transactions*).

One of the examples earlier in this book is an important land tax reform by the Qing dynasty's Yongzheng Emperor, the *tanding rumu*,[10] that incorporated the poll tax into the land tax, with all taxes collected based on the number of *mu* [allotted to peasants, instead of being calculated on how much grain was actually produced]. The *tanding rumu* caused the tax burden to be even heavier on large landowners, but it alleviated the burden on small farmers. As a result, it inhibited the annexation of land and gave more breathing room for small farm owners. Those who have read the historical novel *Yongzheng wangchao* (*The Imperial Court of the Yongzheng Emperor*), by Er Yuehe, probably have an impression about this: in carrying out the *tanding rumu* policy, the Yongzheng Emperor met with fierce opposition from the nobles and aristocrats, even to the point of inducing a crisis in the imperial court. Nevertheless, the implementation of the *tanding rumu* policy revitalized agriculture and enriched the national treasury, and thus laid the foundation for the flourishing of the country under his successor, the Qianlong Emperor.

The economic theories of Alexander Chayanov[11] and Philip C. C. Huang bring to light the reasons for the strong vitality of small farm households. Their theories are somewhat difficult to understand, so I will express them below in simplified fashion, as follows.

---

[9]Wu is one of the eight major language families of Chinese and is spoken in southeastern Jiangsu Province, southern Anhui Province, and much of Zhejiang Province, as well as Shanghai.—trans.

[10]Explained in Chapter 3, Section III, Subsection 3.—trans.

[11]Alexander Chayanov (1888–1937) was a Soviet agrarian economist, scholar of rural sociology, and advocate of agrarian cooperatives.—trans.

In the long sweep of history, under the conditions of too many people and too little land, the labor productivity of small farmers (whether calculated per capita or per unit of time) has been lower than that of large, corporately managed farms. Nonetheless, their average productivity per *mu* of land has been higher than that of corporate farms. This is because, with no other opportunity to make money, small farmers do not care so much about recuperating their expenditures [and therefore can spend great energy in meticulously caring for their crops]. Therefore, [because of the resulting higher yields per *mu*,] small farms are able to compete with corporately managed farms [i.e., large agricultural enterprises].

So what form of management should Chinese agriculture adopt? Why should we insist that farm household operations be taken as the basis? For one thing, with the nation's surfeit of people and paucity of land, the family managed operations would still be an important part of farm production even if capital investments were greatly increased. The economic theories of Alexander Chayanov and the proofs provided by Philip C. C. Huang possess theoretical significance concerning this question. In the following, I will further analyze the situation.

5.3.3.1 The Debate Over Small Farm Economics

In small farm agriculture, there exists a hidden surplus of labor and a surfeit of idle time that conceals the existence of unemployment. These excess laborers are inefficient. Using professional terminology, this idea is known as the "zero value theory of agricultural labor." This economic phenomenon is illustrated by the hidden unemployment theory or the "doctrine of zero farm labor" of Arthur Lewis (1915–1991), a famous economist in the field of economic development. This phenomenon has been long-standing in the agricultural history of China, at least since the Sui and Tang dynasties. From then until right up to the present, there has been a gradual turn for the better only in the wake of the rapid growth of non-agricultural industries.

Nevertheless, Theodore W. Schultz (1902–1998), the famous author of Transforming Traditional Agriculture, *refutes this theory. From his point of view, excess labor does not exist in marketized traditional agriculture. He uses the example of the drop in farm production in India caused by the decrease in agricultural labor that stemmed from the 1918 to 1919 influenza pandemic to prove that in traditional farm production, output and population are closely related: when the agricultural labor force decreases, it is sure to lead to a fall in agricultural output.*

The famous Chinese American economist Philip C. C. Huang refuted Schultz's theory in his scholarly paper "China's small farm economy in the past and present—correct and incorrect aspects of Schultz's theory." According to Huang, the deductions of Schultz fell short of reality in that Schultz hypothesized that all Indian farm households were equally impacted by the influenza pandemic. In reality, that was not the case. The households whose members all fell sick would cause farm yields to drop, but the overwhelming majority of other farm households would still constitute a surplus labor force. Furthermore, if there were no surplus labor force during the busy farming season, it would not mean that there was no surplus labor force at other times.

Many of those in the profession of agricultural economy may not be familiar with Alexander Chayanov, an economist in the former Soviet Union. Nevertheless, I consider him the master of all masters when it comes to research on small farm economy. His work revealed the principle that the organization and the purpose of family farms and enterprise farms are not the same. Under the pressure of too many people and too little land, family farms can continuously invest their labor in the land until marginal returns fall to near zero in order to survive (Lewis' point of view). Nevertheless, corporately managed farms are just like businesses, so when marginal returns are lower than wages on the labor market, they will stop hiring any more workers (Schultz's point of view). Thus, both of these two contending viewpoints have struggled to have the last word.

Philip C. C. Huang provided a further argument. According to him, the yield per mu *(i.e., land productivity) of small family farms (and hence the productivity of small farmers) is higher than that of corporately managed farms [or corporate farmers], but their labor productivity based on calculations of the amount of time worked is lower than corporately operated farms. He called this "involute" or "excess density" production. For this*

reason, family farms are able to bear the burden of relatively low labor rewards and relatively high land rents. Throughout history, family farms have proven more tenaciously vital than corporately managed farms and have even been able to overpower the latter. This also explains why corporately managed farms in the North China Plains during the 1930s cultivated less than 10% of the arable land; small family farms cultivated all the rest.

5.3.4 Small Farm Agriculture Is in a Predicament

China's small farmers, who have relied on intensive, meticulous cultivation of small farms, have consistently supported the world's largest population. But there is no denying that small farms suffer from "High dispersal" and "Low scale"; that they are economically irrational, on the whole; and that they suffer from the "small farm predicament."

I would summarize the state of present-day small farmers and small farms as suffering from the "predicament of low-level development of small farms" (i.e., the "small farm predicament"). Although Chinese agriculture after the rural reforms of the 1980s has made great strides witnessed by the whole world, this problem has persisted. The fundamental reasons for the "small farm predicament" are the strains and growing conflicts found in the relationship between man and the land, as well as the fact that its operational scale is too small. During the Qin and Han periods, a peasant household was able to cultivate fifty or sixty *mu* of land; but to this day, small-scale farm operations still play the major role in China. By the end of 2013, large-production households cultivating fifty *mu* or more had exceeded 2.87 million households, a mere 1% or so of the 260 million farm households. The households that managed one hundred *mu* or more are even fewer.

Based on the differences in agricultural resources between North and South China, the Ministry of Agriculture has determined the standards for major grain producers to be the following: in the south, major grain producers are those farms that cultivate fifty *mu* or more; in the north, it is those that cultivate one hundred *mu* or more. This is the minimum amount of arable land required for rational and economic farm operations as determined by the ministry in charge. But in the wake of the increasingly widespread use of farm machinery and the improvement of service systems, farmers in fact are capable of handling an even greater scale of production.

Small farm agriculture using less than ten *mu* of land per household is lower than the reasonable ratio for people-to-land. This not only leads to a great surplus of labor, but it also makes for low efficiency in production. Because it is under-sized and scattered, small farm production is inefficient; requires high amounts of labor and man-hours; is not conducive to mechanization; and raises the cost of cultivation, harvesting, and application of fertilizer and pesticides, etc. It also limits improvements in efficiency and narrows the room for profits. Taking grain farming households as an example, in 2013 one *mu* of land in a single season earned less than 100 RMB; with an average per household of 7.5 *mu* of cultivated grain, the earnings do not even amount to 1,000 RMB. If the labor force costs were calculated by market value, then those engaged in agriculture would be considered self-employed and their earnings would be considered salaried income.

I believe that China's small farm agriculture has become bogged down in the "small farm predicament"; that is to say, it is "thin-margin agriculture" (even "no-margin agriculture"). Not only is the per-unit area low in profit, total profits are also low due to the small scale of their operations. Low profits subsequently both limit the ability to invest and lower the enthusiasm for investment; this then leads to sluggish application of technology, imprudent land management, and can even result in letting farmland go to waste. Small farms are low in output, efficiency, profits, and investments, and have a low level of land transactions. All of these factors make them difficult to modernize.

Typical reaction to the small farm predicament

On the one hand, small farmers cannot make money, and on the other hand, the price of grain in China in recent years has been higher than that on the international markets. This is a difficult conundrum for small farmers to solve. The predicament has also been intensified by pressure from external markets. Small farmers find themselves unable to "maintain their own integrity" [i.e., using self-reliance to stay immune from these forces], as they did in the age of traditional agriculture.

Now that we have evolved into an industrial society with urbanization continually increasing, commercialization of farm products has become more and more prominent. Apart from those for personal use, the agricultural commodities produced by farmers are for sale on the market; small farms find it impossible to escape the influence of domestic and international markets. In China, they find themselves caught up in a huge global wave of modern development. With production, demand, and prices all being influenced by global agriculture, inevitably production costs, benefits, and efficiency of production should be taken into consideration. However, the "small farm predicament" is like dragging around a huge millstone that hinders the forward progress of modern Chinese agriculture. Can small farms persist? Where is it all heading?

5.4 How to Till the Soil from Now on?—The Family Farm

5.4.1 Family Management Is Still the Main Form

From an international perspective, family management is still the most ubiquitous form of farm management. Modern American agriculture is based on family farms. Its average size farm is around 2,500 *mu*. Even such huge farms are managed primarily by families; of its 2.2 million farms, the great majority are family farms, which make up 86% of the total. The average size of farms in Canada is approximately 4,500 *mu* and, as is the case in America, most are family farms. Farms in European countries are also primarily family farms: 88% of the total in France; 77% in Germany; and an average of 88% for the fifteen European Union countries. Japanese farms have also always been family farms in the main, and for a long time, there have been limitations placed on commercial enterprises engaging in agricultural management.

5.4.1.1 Small Is Beautiful

Small Is Beautiful: *A Study of Economics As If People Mattered*, by E. F. Schumacher (1911–1977), and *The New Peasantries*: *Struggles for Autonomy and Sustainability in an Era of Empire and Globalization*, by the Dutch scholar of rural society Jan Douwe van der Ploeg (1950–), have demonstrated that even in modern Western countries where the economy is relatively developed and farmers are relatively few, family farms form the basis of agriculture.

The New Peasantries divides world agriculture into three modes: peasant; entrepreneurial; and large-scale corporate capitalist (or "food-empire") modes. In my opinion, these divisions by the author are certainly not appropriate. His corporate capitalist mode in fact goes beyond the domain of agriculture. Instead, it describes comprehensive corporate farm production, combining processing, trading, storage, and transport. For example, within the professional field of agricultural economics, one often hears of huge international food companies, such as the American Archer Daniels Midland Company (ADM), Bunge Limited, and Cargill, Inc.; the French Louis Dryfus Company; and the Singaporean Wilmar International, Limited. In fact, they are analogous to China's "leading industrialized agricultural enterprises." Therefore, as far as farm production is concerned, I think that there are just two modes of agriculture: small farm and enterprise.

Although van der Ploeg's divisions are inappropriate, his recognition of small farmers and small farm agriculture is certainly reasonable and enlightening. Van der Ploeg refuted some of the past theories and viewpoints on small farms, including the theories of obstructed development, the theory of withering away, the technical limitation theory, the agricultural involute theory, the theory of impoverishment, and the theory that small farms are lagging behind and are incapable of supporting the world, etc. In his opinion, the intensive labor required of small farms may run counter to economic laws of diminishing returns found in neoclassic economics. Nevertheless, in reality, small farmers' methods of agricultural organization guarantee that their income can remain at an acceptable minimum level or even higher. Small farm agriculture is based on constant utilization of ecological capital (i.e., natural resources), coupled with multifunctionalism. Its labor derives from the family itself (this is the same definition as used in China) or community cooperatives. The land and other production materials belong to the family, and the aims of production are to serve the market and satisfy the needs of the family and of farm production. Small farmers can adopt a number of shrewd strategies to make their agricultural activities steer clear of all those market forces. Consequently, in the process of the realization of small-scale agriculture around the world, developed countries such as Holland, Italy, Peru, and Brazil, have shown a tendency for small farms to reappear. It is necessary to make clear here that the term "small farms" as used in those countries in fact refers to family farms.

In *Small Is Beautiful*, written at the beginning of the 1970s, E. F. Schumacher clearly supported small-scale or intermediate-scale technology (i.e., appropriate technology). He also supported small-scale production and moderate economic units. Aiming at the flaws of capital-intensive and natural resource-intensive economies of developed countries, he pointed out problems of specialization and large-scale production that leads to declines in economic efficiency, environmental pollution, and exhaustion of natural resources. As early as forty years ago, he had already explored these problems in terms of scale of industry and urban social units. At that time, he pointed out that the pursuit of enormous scale was irrational and advocated moderate scale.

Schumacher attached special importance to the largest material resource—land—and its rational utilization. This is not a complicated technical or economic issue, but is fundamentally a philosophical one. He opposed industrialized centralization and specialization of agriculture. He revealed the following: "… large-scale mechanisation and heavy chemicalisation make it impossible to keep man in real touch with living nature; in fact, it supports all the most dangerous modern tendencies of violence, alienation, and environmental destruction. Health, beauty and permanence are hardly even respectable subjects for discussion…."[12]

In Chap. 4 of his book, Schumacher also proposed the fascinating concept of "Buddhist Economics," which explores how to achieve pre-set goals with the least means and how to enjoy the greatest satisfaction through the best methods of consumption. It is a very different concept from modern utilitarianism economics, which advocates expanding consumption as the best way to promote production.

In contemporary China, all walks of life have accepted that "big is good and small is bad." In commerce and industry, large enterprises, big projects, and massive production or management are worshipped. In the competition to increase GDP, local governments race to introduce all kinds of large industrial enterprises,

---

[12]E. F. Schumacher, *Small Is Beautiful: A Study of Economics As If People Mattered*, Chapter 7.—trans.

regardless of associated problems of environmental and ecological protection. In the development of towns and cities, there is also the pursuit of bigger and newer. Even in agriculture, "corporatization of agriculture" has been proposed as the slogan under which to carry out massive corporatization-style production by buying up huge numbers of small farms and consolidating their land into enormous tracts.

Why do all enterprises spare no efforts to pursue the expansion of capacity? Why cannot there be a slowdown in the expansion of large numbers of industries that are highly polluting and that consume high levels of energy? Why are small and midsized enterprises never able to thrive? For what reasons are our cities still spreading out like giant pancakes on a grill? All of these questions are related to our habits of consciousness. Schumacher's "Small is beautiful" and van der Ploeg's "A world in which small farmers exist is more beautiful than one without them" enlighten us to reflect philosophically on problems of economic scale and development style.

5.4.2 *Management Scale Needs Expanding*

The household contracted management system is an important part of the basic management system of rural areas in our country and is the cornerstone of the CPC's rural policy. Since the rural reform, the implementation of the household contracted land management system has motivated a wide range of farmers and has liberated the productive force of the rural areas and promoted the great development of the rural economy. Practice demonstrates that implementation of household contracted management is the historical choice of great numbers of farmers, is in accord with the basic national conditions of China, and possesses broad adaptability and exuberant vitality. National policy emphasizes repeatedly the unswerving adoption of a two-tier management system with a combination of centralization and decentralization, with household contracted management as the basis.

Nevertheless, in wake of the transformations of farm production and the increase in the level of productive forces, the household contracted management system has begun to present certain inadequacies. The main problem is the small scale of family management, with the salient features being the "small farm predicament" (which was explained previously in this chapter). That is because modern farm development requires some certain scale of management size for efficient production. At the same time, the labor force in rural areas is in the process of migrating away (especially young people in the prime of their lives). "Who will till the land?" (i.e., "Who will operate the farms?") has become a very big question. With the "small farm predicament" and the outflow of labor from the countryside, is the household contracted management system still needed? The answer is yes.

In recent years, the rural farm economy and population structure have undergone major changes. Over 260 million farm laborers have migrated to non-agricultural industries. Furthermore, the nation-wide rate of urbanization has risen year after year and has now already passed the 50% mark. Transactions of farmland have been normalized and the rate of transactions has been increasing. In some areas, farmland is even being abandoned and reverting to wasteland.

Despite these facts, the household contracted land management system should still be adhered to. China in total has 260 million farm households. On average, there is only about eight *mu* per household. The national situation is that there are many farmers and farmland is scarce; hence, it is impossible in the long run for the majority of farm areas to realize suitable-scale management by expanding land holdings to achieve economies of scale. Therefore, it is simply unrealistic to abandon the present household contracted land management system.

The transformation of farmers into city residents is a long-term process, and the shift of farmers to nonagricultural industries is still in a precarious state. A great majority of migrant workers as yet do not enjoy full urban benefits, such as housing, health care, education, and social welfare. In addition, some years later the overwhelming majority of these may have no choice but to return to rural areas to farm. If the problem of unemployment caused by farmers not being able to find work in nonagricultural jobs is not solved, there will be undue stress placed on the development of large-scale farm management. Farmers will then simply continue to be pushed toward the cities; this in turn would cause a great many farmers to lose their land and fall into extremities. It would also create large numbers of urban or rural poor, bringing about serious social problems. Therefore, stabilizing the dominant position of farm household management is important for the stable development of agricultural production, rural areas, and society as a whole.

Adhering to the household contracted management system is not simply maintaining traditional small farms. As our country pushes forward industrialization, urbanization, marketization, and informantionialization, developing modern agriculture also requires constant refinement of the basic rural management system, i.e., the perfection of the household contracted land management system. An effective, properly scaled, and new model of the family farm management system is the most basic institutional arrangement to safeguard farmers' livelihoods in a country with so many people and so little land. Likewise, family-farm management can take extensive advantage of all sorts of modern production methods, and even utilize social division of labor[13] to put its own production on the track of mechanization, specialization, and commercialization. The use of improved seed strains, chemical fertilizers, pesticides, and irrigation technology, etc., is rarely affected by the scale of farm households; but in regard to the use of agricultural machinery, small-scale farm households will admittedly encounter some limitations. Yet, by purchasing appropriate farm machinery or paying for socialized [i.e., cooperative] services of such machinery, they can attain the same high level of mechanization and efficiency. Family farm management can also integrate with cooperative management and professional management to improve its economic efficiency.

[13]The term "social division of labor" is Karl Marx's contribution to division of labor theory and is in contradistinction to "economic division of labor." Used here, it is a neutral term and implies using various (unspecified) groups within the family or community to get needed work done—trans.

Through undertaking land transactions, it can realize properly scaled management of family farms and extend implications of what the family farm means. By improving the external environment of small farms, it can achieve mechanization, professionalization, and socialization of related sectors and overcome shortcomings of decentralized management of households, strengthen family farm management, and further develop agricultural production.

The way out of the predicament for small farmers is not restricted to one model; there can be several methods, such as migrating to do manual work, operating a business, or running a factory. However, these are nonagricultural channels. With regard to managing agriculture, the number-one prerequisite to enable a small farm to escape its predicament is to enlarge the scale of its production. A single small farmer who is able to plant only 10 *mu* of corn certainly will not be an economically happy farmer. Our household once farmed less than ten *mu*. How could such a small plot of land reasonably support our whole family?

Suitably scaled management is the objective requirement for modern agricultural development. Modern farming requires extensive, large production and intensive management. With such production, machines can be used effectively; with intensive management, costs can be reduced and labor productivity can be increased.

Research clearly shows that the small-scale machinery level and the appropriately large scale of household management (using family members as the major labor force, mainly producing grain) that are achievable in China both require managing four to five times as much land as at present—i.e., a minimum of approximately forty *mu*.

This is in basic accordance with the data we collected in 2011 from a survey of 669 farm households across the country on the subject of farmland transactions. Those farm households believed that under existing conditions the average scale for appropriate land management should be 33.9 *mu*, but that with improved production conditions, the suitable size on average could be 102.5 *mu*. Ministry of Agriculture survey data shows that, considering the natural resource conditions and the income gap between urban and rural areas of China, economies of scale can be achieved if, on average, the household areas that can grow two crops a year cultivate fifty to sixty *mu* and those areas that can only grow one crop cultivate 100 to 120 *mu*. According to research of UNFAO, the agricultural scale that is internationally competitive for growing cash crops is no fewer than 170 *mu*. For growing grain crops, it is no fewer than 300 *mu*.

5.4.3 People Who Manage Agriculture Should Be Professional Farmers

Small-scale management is the major characteristic of small farms. It is also the main reason for the "small farm predicament." Once the scale of farm household management reaches a rational level, with its manpower matching its management ability, small farmers can grow and become professional farmers and manage family farms (or even become large, specialized farmers and breeders). By then, their operational conditions will have taken a turn for the better and their income will have increased to be equivalent to that found in nonagricultural employment. For a period of time in the future, the main form of agricultural operations will be family farms and small farms, but only moderate-size-or-larger family farms that are managed by professional farmers can be the vehicle to carry Chinese agriculture out of its predicament and in the direction of transforming small farms.

What needs to be emphasized here is that the process of small farms moving toward transformation into family farms has to be step-by-step and in an orderly fashion. We cannot just coerce or induce farmers to transact their land merely for the sake of pursuing an increased scale of farm management; we cannot piece together family farms simply by confiscating or lumping together land in the fashion that large producers have done in the past.

Currently, China possesses conditions to achieve adequately large-scale management. The prerequisite for this is to shift part of the farm labor force to secondary or tertiary industries in order to ensure that those workers obtain full employment in nonagricultural industries. Generally speaking, countries around the world have transferred, step-by-step, 60–70% of their farm labor into nonagricultural industries as an initial condition for implementing agricultural operations on a large enough size to achieve efficiency and reap the benefits of economies of scale. Along with industrialization, with a resultant great shift in the rural population and together with the promotion of the market economy, most developed countries have achieved large-scale farm management by gradually concentrating land in fewer-but-larger farms, in addition to taking other steps to achieve agricultural modernization. As detailed below, China already possesses certain economic and social conditions needed to implement appropriately large farm operations, so that these agricultural changes are inevitable.

At present, China is in the middle stages of industrialization and is in an accelerating period of urbanization. In 2013, the ratio of value to output of Chinese agriculture was only 10%, while for primary industry and secondary industries it was 44%. The extent of urbanization exceeded 53%. In the wake of the increasing absorption of rural labor by cities and industry, it is more and more clear that the proportion of the rural labor force engaged in farming has steadily declined.

Of the 269 million farmers, the new generation of migrant workers who were born after 1980 totals 125 million. They basically do not know how to farm and in the future would not be able to make a living at it. In the long run, they are likely to become alienated from farming and the land. At the moment, conditions for properly large land management already pertain in developed areas along China's eastern seaboard and in the suburbs of some of the large and medium-sized cities, as well as some midwestern regions from where large numbers of laborers have migrated.

By the end of 2013, transactions of contracted farmland in China had already reached 340 million *mu*, which accounted for 26% of the total nationwide contracted cultivated area. There were over 2.87 million large-scale specialized producers, each of whom managed at least 50 *mu* of land. The number of family farms exceeded 870,000. It is evident that the total amount of farmland held by households that enjoy large-scale management is relatively little and is insufficient. Yet, in some developed areas, through farmland transactions that increased land holdings in order to make large farms, large-scale farm management has been achieved.

"Family farm" is not a new term, but since 2013 when the CPC Central Committee's *No. 1 Central Document* took the family farm as the new main form of agricultural management, "family farm" has become a trendy buzzword in agriculture throughout China.

In February, 2014, the Ministry of Agriculture issued *A Guiding Opinion Relating to the Promotion of Family Farm Development* that gave a relatively long definition of the term "family farm":

> The family farm is regarded as the primary new body of agricultural management. Members of farm families are the main source of labor, and income from agriculture is its main source of income. It utilizes the family as the entity for contracting or transacting land. It is engaged in large-scale, intensive, commercialized agricultural production. It preserves farm household management as its core, and upholds the basic position of family management. It suits our nation's basic conditions, and accords with special features of farm production and with social and economic developmental stages. It is an updated version of farm household contract management, and has already effectively taken the lead in appropriately large-scaled management and the development of modern agriculture.

At the same time, the document indicated that there are three basic, distinctive features of family farms:[14] First, family farms primarily depend on family labor rather than on hired labor. Second, family farms specialize in agriculture, principally carrying out specialized cultivation and breeding, with most of the operators having received agricultural education or vocational training and possessing the ability to produce farm commodities and goods. Third, family farm management scale is appropriately large, and the income level is commensurate with that of residents of local cities and towns, achieving relatively high rates of land yields, labor productivity, and utilization of resources. Another concept of the family farm which is commonly quoted is that it is the new main form of farm management, one that uses family members as the main source of labor to engage in extensive, intensive, and commercialized production and management, and uses the income from farming as the family's main source of income.

What is the present nationwide situation of family farms? According to statistics from the Ministry of Agriculture, by the end of 2012, in the nation's thirty provinces, regions, and special municipalities (Tibet and Taiwan not included), there

[14]At the beginning of the 1970s, America proposed three distinctive conditions as the standard for what constitutes a family farm: first, family members operate and manage the farm; second, families assume the risks; and third, families must provide at least half of the labor on the farm. As I understand it, only family farms are able to enjoy preferential government policies.

were altogether 877,000 family farms that met the conditions to be surveyed. The arable land that they managed totaled 176 million *mu*, with an average operational size of 200 *mu*, constituting 13.4% of contracted arable land nationwide. The average labor force of family farms was 6.01 persons; of these, an average of 4.33 were family members and 1.68 were long-term employees.

The Ministry of Agriculture data also showed that, in 2012, nationwide income of family-managed farms totaled 162 billion RMB, with an average family farm income of 184,700 RMB. In some agricultural industries, family farms have started to play important roles. At the beginning of 2013, the Crop Farming Bureau under the Ministry of Agriculture issued results of a survey showing that there were 682,000 households that were large producers of grain, making up 0.28% of total farm households. These households cultivated an area of 134 million *mu* and their land constituted 7.3% of the total farmland nationwide. Their grain yields reached 149.2 billion *jin* [i.e., 76.6 billion kilograms] and made up 12.7% of the total grain output of the country.

How do the standards for small farm management and family farms differ? The difference lies in the scale of their operations. At the beginning of Section III in this chapter, the small farm was defined as an individual farm household that totally or mainly relies on its own labor to operate farming on a scale that is within its ability. The labor source of both small farms and family farms primarily comes from the family itself; they are both engaged in farm management, so there is no difference on these two points. The only main difference lies in their scale. If the scale of a household-managed farm is smaller than its labor and management ability, then it is a small farm. If its scale is greater than its labor and management ability, then it is a family farm. Of course, labor and management ability vary from person to person: some people have more such ability; some have less.

Moreover, China is a vast country; natural resources and economic and social circumstances in each region vary because of differences in soil fertility, precipitation, intensity of sunlight, accumulated temperature, number of frost-free days, level of farm mechanization, labor availability, etc. In this sense, it is extremely difficult to set up unified and strict criteria.

However, eventually it is still necessary to set up criteria that are general, in accordance with national conditions, with an economically appropriate scale, and that conform to societal standards. Take grain production as an example: according to present production conditions, a sown area between 100 and 120 *mu* is the standard cutoff point for differentiating a small farm from a family farm. With this scope or larger, there is a high allocation efficiency of production elements such as labor force and land; land productivity, labor productivity, and utilization of resources are also high. Family farmers' incomes are about on par with non-agricultural income levels. However, if the scope is smaller than this, one then runs into the "small farm predicament" noted earlier.

Income from family farms is greater than hiring out as manual labor

Based on the above Ministry of Agriculture statistical data, 99% of Chinese farm households are small farms and less than 1% are categorized as family farms; but the latter cultivate 13.4% of China's contracted land, with an average size of 200 *mu*. In addition to grain production that differentiates small and family farms based on sown areas, there still needs to be differentiating criteria based on capital investment, labor investment, and the value of agricultural products such as vegetables, fruit, tea, livestock breeding, etc.

For example, in the case of the output and profit of one *mu* of vegetables grown in a large greenhouse,[15] the income is higher than growing ten *mu* of grain; but the investment needed for vegetables grown in a large greenhouse is also higher and the labor required is much more. Therefore, in this case the criteria for differentiating a small farm from a family farm are both complex and different. The criteria for large-scale management in Suzhou, Jiangsu Province are: for large crops such as grain, cotton, and oilseed, fifty *mu*; vegetables grown in open fields, thirty *mu*; fruit, thirty *mu*; aquatic products, fifty *mu*, etc. One can regard these as the criteria for differentiating small farms and family farms in various kinds of agricultural industries.

Prior to the CPC Central Committee's *No. 1 Central Document* of 2013, there was no concept of the family farm in official policies, but rather other terms were used such as "large-scale crop and livestock producers," "large specialty producers," and "large grain producers," etc. In essence, there was not much difference made between family farms and such large crop and livestock producers; they were basically considered the same thing. But in the past, there was no specific definition on crop and livestock producers; their concept and criteria were relatively vague.

[15]In China, these are typically long, temporary, Quonset-like shelters, made of arched frameworks of bamboo or tubing of plastic or metal and covered with transparent plastic sheeting.—trans.

The term "family farm" is now used officially. Moreover, guiding policies have been issued to define the concept, establish criteria, and regulate administration. Suffice to say, the term "family farm" is an updated version of the old term, "crop and livestock producer."

At present, there is some misunderstanding in knowledge and in practice. In 2013, I visited a so-called "family farm" that in reality possessed salient features that were more in keeping with those of a farmers' professional cooperative. There were scores of farm households whose members had become equity investors in a farm made up of contracted land. The farm owner had in advance paid a guaranteed minimum rent of 200 RMB per *mu* to the holders of the rights of the contracted land. He had furthermore provided each farm household with production materials, such as seeds, fertilizer, and pesticides. The farmers were responsible for field management. The farm owner then sold all the vegetables they produced, and then paid out profits to the farmers in the form of dividends. After talking with the farm owner a while, it was clear he did not really know what a family farm was. According to the definition of the Ministry of Agriculture, this farm would not be considered a family farm, because it had the distinctive features of both a cooperative and an agricultural enterprise (i.e., a farm operated as a business).

From surveys conducted by scholars and the media, it is clear that there have been similar situations with other so-called "family farms." Cooperatives or company-managed farms are registered as family farms in anticipation of preferential government policies that will support family farms. Family farms, farmers' professional cooperatives, and farm enterprises are obviously not the same. Farmers' professional cooperatives are cooperative economic organizations. Agricultural enterprises are also not based on management by farm households. Nevertheless, some local governments have no clear understanding of this and are anxious to pursue change and success; they blur the boundaries between family farms, cooperatives, agricultural enterprises, and other such kinds of management.

5.4.4 National Policy Should also Be Supportive

Family farms managed by professional farmers are a major form of future agricultural production and management. The government needs to strengthen the strategy of training of professional farmers; carry out a supportive strategy for family farms; and, in a multi-faceted manner, transform small farmers into professional farmers to manage family farms. Government can thereby lay the management foundation for farm modernization.

In 1987 in *No. 5 Central Document*, the CPC Central Committee for the first time clearly put forward the policy to adopt different forms to carry out appropriately large-scaled management. The following related policy papers all mentioned this kind of development: the eleven *No. 1 Central Documents* since the 21st century, the Decisions from the Third Plenary Sessions of the 17th Central Committee; the *Decisions* of the Third Plenary Sessions of the 18th Central

Committee; and the *Development of a Nationwide Agricultural Economy*, part of the 12*th Five-Year Plan.*[16] It can be concluded that this has already become the future direction of agricultural management. Ever since the term "family farm" was proposed for the first time in the 2013 Central Committee's *No. 1 Central Document*, to develop family farms based on suitably sized management has become an important choice among the main new forms of farm management.

In recent years, some areas such as Jiangxi, Jiangsu, and Hunan provinces, have issued specialized subsidy policies to promote sizable operations. Zhangjiagang, a county-level city in Jiangsu Providence, launched a land-transfer subsidy system in 2007, offering farm households an annual 300 RMB per *mu* for giving up control of their land; afterwards, all of the cities and districts in Jiangsu followed suit. Since 2011, Suzhou[17] has spent 300 to 400 million RMB every year on land transfer subsidies. This has effectively promoted amply large-scale farm management, which has reached 91% of the farms in this district.

At present, at the national level there is still a shortage of concrete action to implement large-scale management of agriculture. This is perhaps out of concern for situations such as: land expropriations; local governments or business enterprises taking advantage of farmers by violating their land rights and interests; government operational costs; and the problems of precisely targeting subsidies, etc. Moreover, some areas for several years have already implemented subsidies targeting land transactions and large-scale agricultural management; under these circumstances, the central authorities can learn from local experience and draft specialized guidelines that promote large-scale management. They can also encourage local governments to support large-scale management and family farms under the conditions of adhering to basic rural management systems and protecting farmers' land rights.

The subsidy policies for family farms and small farms (ordinary farm households), should be applied fairly, but should favor small farms that can mature into family farms. National farm subsidies and development projects should set up special programs to support operations to help them reap the benefits of economies of scale. In addition to other departments, those government departments dealing with central finance, agriculture, and comprehensive agricultural development, should lead and support large-scale management and family farms on the basis of current policies and projects. Through policy support, they should put an end to "the dilemma of low-level development of small farms" and guide large-scale management, creating the basic conditions for achieving agricultural modernization.

[16]I.e., the PRC economic plan for 2011–2015.—trans.

[17]Suzhou is an administrative district that is composed of a much larger area than just Suzhou City itself, and includes Zhangjiagang and four other county-level cities and rural environs.—trans.

5.4.5 Supporting Services Need to Keep Pace

In the process of small farms maturing into family farms, farm households will not be able to be well developed if they depend solely on their own production and management of family farms. In China, family farms cannot leave the protective services provided by collectives, cooperatives, and society. Service organizations such as collective economic organizations of rural areas, village committees, farmers' professional cooperatives, and agricultural and social services are guardian angels of the healthy development of small and family farms.

China's rural areas have implemented a system of collective ownership of land and other property. The farmers as a whole are the collective owners of all this. The collective economic organizations or rural committees are the agents for the property and collective economy in rural areas. They are the organizers of cultural infrastructure, economies of scale, and socialized mass production; furthermore, they are the foundation for village regulation. These collective economic organizations need to be brought into full play in unifying management and service. They need to be guided to do the following: utilize capital, property, and specialty contracts; develop equity investment operations with large contractors and skilled professionals to resolve, matters household by household, those things that cannot be handled well or at all by households; and promote the healthy development of family farms in order to better serve household management.

Farmers' professional cooperatives are cooperative economic organizations. They have the power to concentrate, condense, or distribute the strength of farm households; integrate and optimize production and management; and provide farmers who have joined the organization with preproduction, midproduction, and postproduction services. There should be support for the development of cooperatives through financial investments, preferential taxation, and financial services, etc., to increase the service level and ability of cooperatives relating to farm technology, information, credit, processing, marketing, etc.

Under the conditions of the market economy, agricultural social service organizations play an increasingly important role. Therefore, we should do the following: Accelerate the education provided to various kinds of social service organizations. Provide social services for family farms, such as standardized supplies of seedlings, technological standards, and specialized services. Bring the key links of family farm production and management to within the scope of socialized farm services. Provide family farms a social service system that combines public benefit services with business services and that coordinates specialized services with comprehensive ones. We should also encourage various kinds of service organizations and family farms to establish closely connected mechanisms of mutual interest, bringing those mechanisms into full play in providing unified management and services for small farms and family farms.

Conclusion
The Future of Agriculture

"Ten High" Agriculture and "Ten Top Strategies" for Agriculture

In the August 20, 2013, issue of the *Nongmin ribao* (*Farmers' Daily*) I published an article, ("How to respond to the challenges of 'Ten High' agriculture—phasic characteristics and policy orientation of Chinese agricultural development"). It summarized Chinese agriculture's "Ten high" features and advanced some relevant directions for policies:

"High degree of aging" of the farm labor force means that we need to make great efforts to train professional farmers. "High degree of decentralization" means that we should actively develop suitably scaled management. In addition, "High costs" of farm production means we need to relieve cost pressures through multiple channels. Agriculture has entered a period of financial "High investment," so we still need to steadily increase investment and subsidies. Agricultural development depends on a "High degree of science and technology," so we should go further in promoting agronomical innovation and application. "High pressure" in food security coexists with "High import" of farm commodities, so we must use an overall plan to stabilize our domestic production base and expand our international markets. "High consumption" and "High pollution" present us with a grim situation, so it is imperative to transform methods of agricultural development. Finally, "High risk" in agriculture is becoming evident, so we need to strengthen the risk management system to protect modern farming.

The above-mentioned "Ten High" features of agriculture are from the perspective of problems. The corresponding policy orientations are in fact solutions and future directions: professional farmers, large-scale management, scientific and technological innovation, food security and safety, sustainable agriculture, and national financial support—all are ways to solve the problems confronting agriculture's future.

Y. Zhang, *Insights into Chinese Agriculture*,
https://doi.org/10.1007/978-981-13-1050-8

Afterwards, I summarized in "Ten Top Strategies" policies for agricultural development: conduct professional farmer training; large-scaled management; regulate costs and prices; provide financial investment support; drive for scientific and technological innovation; attain grain self-sufficiency; promote agriculture's overseas investment and marketing; ensure sustainable development; pursue quality and safety of farm products; and provide for risk management and prevention.

In other words, the article's title could have been "Chinese Agriculture's Problems, Solutions, and Future."

After the article was published, it had some influence in the field; for example, the Zhonghua Fertilizer Company even requested that its workers and staff study it. Reactions to it encouraged me to write a book on agriculture that was both professional and easy to understand. I took the "Ten Highs" as a starting point; restructured it in a fresh way; added substantial content; and searched for data. I spent an intermittent two years to complete it while working in Guangyuan in Sichuan Province (temporarily reassigned there from my original work unit) and in Beijing. As the manuscript was about to go to press, I summarized the "Ten Lows" to contrast them with the "Ten Highs."

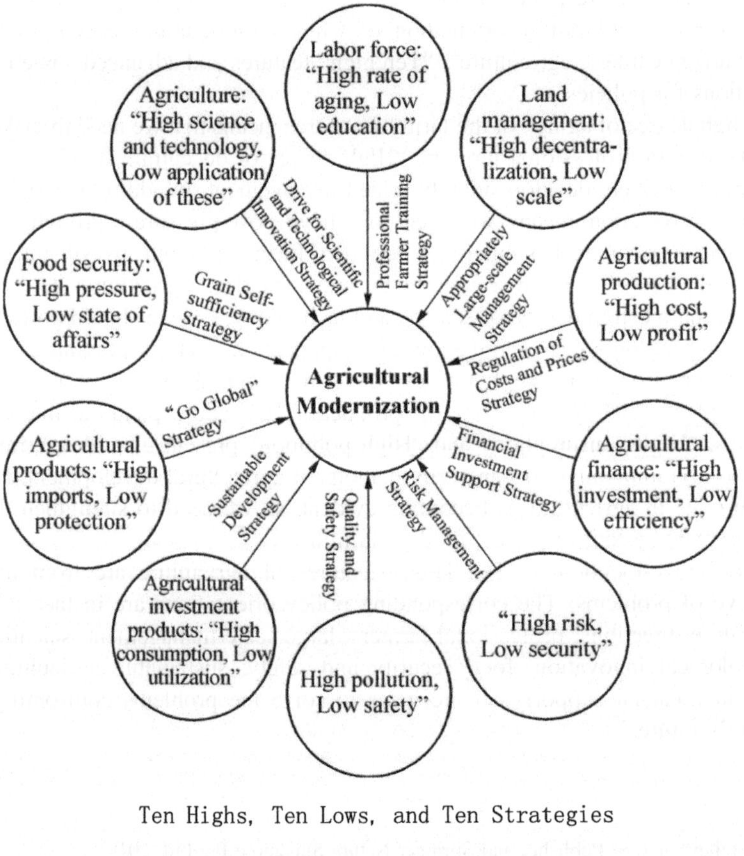

Ten Highs, Ten Lows, and Ten Strategies

To sum up: The labor force is "High in aging, Low in education." Land management is "High in decentralization, Low in scale." Production is "High in costs, Low in profit." Finance is "High in investment, Low in efficiency." Agriculture is "High in science and technology, Low in application of these." Food security is "High in pressure, Low in its actual condition." Commodities are "High in imports, Low in protection from imports." Investment inputs are "High in consumption, Low in efficiency of utilization." In addition, there is "High pollution, Low safety" and "High risk, Low security." In light of these "Ten Highs" and "Ten Lows," only by implementing the aforementioned "Ten Top Strategies" can we achieve agricultural modernization.

Characteristics of Agriculture in Different Areas: "Let a Hundred Flowers Bloom"[1] in Agriculture

Those living in the mountains, live off the mountains; those living near the water live, off the water.[2]

China is a vast country, with each area having its own unique agricultural environment, distinctive regional features, and relative advantages. The agricultural industries, production, and lifestyles that farmers from different areas rely on to survive and thrive are not the same. Agriculture and its characteristics varying according to regions—so let a hundred flowers bloom when discussing its special distinguishing features. The past, present, and future of farming in each area is not the same, either.

Northwest China has vast, endless grasslands; the grasslands of the herdsmen of the provinces of Inner Mongolia, Qinghai, Xinjiang, and Tibet are measured in the millions of *mu*. The meadows are not small, but because these areas suffer from shortages of water and drought all year long or are frigidly cold due to high altitudes, the land can only be used for growing grass to graze livestock. The supply of mutton of these pastoral areas is abundant, but they rely on other places to provide them with grain, vegetables, and tea, with the prices for them higher than they are in the cities. In 2007 and 2009, I spent a total of two months in these pastoral areas conducting research. I was often amazed that a lot of fellows there were able to eat two large platters of mutton in one sitting, but vegetables were a rarity on the table. According to the local cadres, because the food there is monotonous and the climate tends to be cold, the herdsmen of the grasslands are susceptible to malnutrition, gastroenteritis, arthritis, and other such maladies.

The black soil of the Northeast seems boundless in all directions and constitutes one of China's great granaries. In Heilongjiang Province, many farm households operate on over one hundred *mu* of land, a scale that is much larger than that of other agricultural areas, and the land is suitable for mechanized farming. However, the Northeast has very low cumulative temperatures and a very limited frost-free period; only one crop per year can be grown.

---

[1]I.e., "Let a multitude of different views be expressed."—trans.

[2]This is a common Chinese saying, meaning that one must make a living according to one's given circumstances. —trans.

In China's southwest area, Sichuan, Guizhou, Yunnan provinces and Chongqing (a provincial level metropolis that includes a large hinterland) are hilly and mountainous and difficult to access, so in an economic sense there are a lot of poverty-stricken people. Nevertheless, Heaven is always fair. Just as when one door closes a window is then opened, so in compensation, in the southwestern region the rainfall is abundant, the weather warm, and the flora luxuriant. All year around is like a warm awakening of spring, and it seems that all a farmer has to do is scatter seeds on the ground for them to sprout. I am incessantly envious, since I come from the steep, loess slopes of Shanxi Province, where we suffer drought nine-out-of-ten years. In the Southwest, all sorts of vegetables, fruits, and meat are especially abundant and delicious; even in winter, one can eat fresh pea shoots and Chinese broccoli that are not grown in greenhouses, and in early spring one can savor fresh cherries and loquats.

In Jiangsu and Zhejiang provinces on the lower reaches of the Yangtze River, Suzhou and Hangzhou are heaven on earth, and home to a cornucopia of fish and rice. What accounts for the numerous talented people that historically have been produced by this region south of the Yangtze River? It is simply because of the natural resources of the land. Southern Jiangsu Province together with Zhejiang, Anhui, Hunan, and Hubei provinces form a large farm belt that has comprehensive advantages: weather is moist and mild, terrain is generally flat and accessible, river systems flow all over the countryside, and soil can grow grain, vegetables, and fruit in great abundance. Fish and freshwater shrimp and crabs are available everywhere in rivers and lakes. Therefore, since ancient times, this region has been a rich and prosperous land. Its agricultural advantages and abundant farm products have promoted development of education, science and technology, and the flourishing of handicrafts, commerce, and industry.

Step-by-Step Agricultural Modernization

Modernization is the objective of present-day agriculture, just as it was for traditional agriculture. Nevertheless, agricultural modernity certainly has no specific form and is not static. Modernization is both dynamic and gradual, and its implications are endlessly rich. Farming using plows pulled by oxen is modern compared to slash-and-burn, but today they both seem traditional and backwards.

Modern Chinese agriculture at one time could have been summed up in six words: high productivity, high quality, high efficiency. However, because of the increasingly apparent ecological, environmental, and food safety problems, implications of up-to-date agriculture have to include ecology and safety. A perfect balance needs to be maintained between the factors of high production, high efficiency, high quality, ecology, and food safety. One cannot merely seek high production and high efficiency while at the same time harming the ecosystem and food safety; conversely, one cannot merely talk about the environment and food safety without regard to basic production output and food demand.

Although symbolizing contemporary agriculture, mechanization, chemical fertilizers, pesticides, large concentrations of animal breeding, and even transgenic technology, are not really 100% beneficial with no harmful aspects. For example, there is abuse of chemical fertilizers and pesticides; high consumption of energy brought about by farm mechanization; difficult problems of epidemic diseases and the disposal of polluting wastes caused by large concentrations of livestock; and safety questions about genetically modified farm products.

Pursuing modernity and maintaining objective balance requires governmental support, guidance, and supervision; even more than that, it requires great efforts and struggle of farmers, agriculture-related enterprises, and agricultural cooperatives, etc. Only in this way can the brilliant 5,000 years of agricultural history of the Chinese people continue, and China's position as an agriculturally great and powerful nation be consolidated.

The future of agriculture is already in route, right under our feet.

Epilogue

For the last few years, it has been my dream to write a book on agriculture that is both popular and at the same time professional. This idea originated from the fact that although today there are fewer and fewer people directly practicing the profession of agriculture, at the same time there are more and more people paying close attention to and getting involved with farming. On the other hand, most books on agriculture are too specialized; it is rather difficult for the nonspecialist to read them. As a result, people have a blurry sense about it. In addition, there are misreadings and misunderstandings about its policies. For these reasons, as a policy researcher specializing in rural economy, I began attempts to make this speciality easier to be understood by general readers.

My 2013 article on the "Ten Highs" of agriculture was the embryo of this book. For the next year and a half, I intermittently wrote different chapters and finally finished the first draft. It was originally organized along the lines of the Ten Characteristics of the present stage of our country's agriculture and Ten Strategies on how to deal with it. Kuang Zhihong 匡志宏, Deputy Chief Editor of the Shanghai Far East Publishing Company, read it and thought that I did not need to follow the usual formula for books dealing with agricultural economy; since this book is supposed to be a popularized version, I could integrate popular content in order to make it more readable. Under his suggestion, the manuscript framework was thus finalized as five chapters together with an introduction and conclusion.

After the first draft came out, I invited some expert scholars and friends to make comments. I am extremely thankful for their patience in reading it and proffering such cogent suggestions. Among these, there were my colleagues: researchers Cui Chuanyi 崔传义, Cui Xiaoli 崔晓黎, Xiao Junyan 肖俊彦, and Zhou Qunli 周群力, Ph.D. Furthermore, there were: Luo Dan 罗丹, Ph.D., from the Rural Work Leading Group Office of the Central Committee of the CPC; Dou Penghui 窦鹏辉, Ph.D., from the Ministry of Agriculture's Scientific Education Department; Yang Xiaoyan 杨晓艳, Ph.D., from the Ministry of Land and Resources, Land Restoration Center; Liu Tongshan 刘同山, Ph.D., from the School of Agricultural

Y. Zhang, *Insights into Chinese Agriculture*,
https://doi.org/10.1007/978-981-13-1050-8

Economics and Rural Development in Renmin University of China; and Zhao Jianping 赵剑萍, from Dadi Hanke (Chengdu) Biological Technology, Ltd. In addition, from the Shanghai Far East Publishers there were: Deputy Chief Editor Kuang Zhihong 匡志宏, Editor Li Ying 李英, Associate Senior Editor Yang Lincheng 杨林成, and Editor Qi Dongcheng 祁东城. All of the above proposed many valuable suggestions.

I want to especially thank the editors and specialists at the Shanghai Far East Publishing Company for their professional and detailed work. On the basis of their critical comments, I spent another six months of my spare time supplementing and improving the contents of the manuscript.

My Ph.D. supervisor, Professor Kong Xiangzhi 孔祥智, of the School of Agricultural Economics and Rural Development in Renmin University of China, made use of days off during the Asia-Pacific Economic Cooperation (APEC) Conference in Beijing [in November, 2014] to write an elegant and beautiful preface to this book. I especially appreciate Professor Kong's concern and care for me over the years.

I have been working in the field of rural economy for over ten years, during which Han Jun 韩俊[3] has always guided my research efforts with the utmost care. He has always assiduously corrected reports that I have submitted. I thank him from the bottom of my heart for opening up my research horizons, helping me to delve into specialized realms, and urging me to elevate my writing ability. I also thank him for taking time from his busy schedule to write a preface for this book from his strategically advantageous perspective.

In rural economy research work, my leaders, colleagues, and many other fellow researchers from the Research Department of Rural Economy of the Development Research Center of the State Council, including Ye Xingqing 叶兴庆, Director of Rural Department; Liu Shouying 刘守英, Deputy Director; and He Yupeng 何宇鹏, Deputy Director, have given me great support and assistance. I want to express my deep appreciation to them all.

Around thirty years ago, during my boyhood, on weekends and holidays I would follow my father and mother into the fields to plow, hoe millet, or dig potatoes. Farm work seems easy; but in practice, it is both difficult and bitter. After a couple of hours of work, my hands and feet would go numb. At that point, my mother would lecture me:

Study hard, get into a good university, and then there will be no need to browbeat the clods of earth in order to make a living.

Nowadays, although I no longer struggle with the earthen clods on the farm, the indissoluble bonds between agriculture and me have become ever stronger.

Writing about farming is similar to farming: there is plowing, sowing of seeds, and weeding. Only then come the blooms and finally the harvest. Over the past ten

[3]Han Jun was originally the Deputy Director of the Development Research Center of the State Council and is now the Director of the Rural Work Leading Group Office of the Central Committee of the CPC.

years, the more I have investigated agriculture and pondered over it, the more anxious and worried I have become; this is all reflected in this book.

Agricultural problems will certainly not be easily resolved in the wake of the advancements of the times. On the contrary, new circumstances of agricultural globalization and increasing eco-environmental pressures will cause problems to become even more complex. The agricultural industry's perpetual theme is to not only accelerate development and increase the income of farmers, but also to guarantee that people can obtain sufficient and safe food. In order to develop well, agriculture depends not only on the hard toil of farmers, but also on the attention and protection of the whole of society.

To understand farming is to feel a sense of longing for one's native place. Humanity for the last 10,000 years has fostered farming genes that are now deep in our marrow. Agricultural civilization that has developed through several thousand years of humanity is deeply rooted in our hearts. Whether you or I are farmers or whether we work in agriculture, we all to some extent have feelings toward farming. We all actually or metaphorically have a piece of heaven and earth waiting to be discovered, understood, cultivated, and harvested.

In closing, I would like to say that even though one wishes to cultivate good soil, one is not necessarily able to do so; in the same way, even though one wishes to write a good book, one is not necessarily able to do so. Therefore, because I have little talent, even less learning, and display a lack of progress in my studies, I respectfully request the reader's understanding and ask for corrections for any mistakes of omission or commission or for any inappropriate, obscure, or obdurate passages in this book.

Zhang Yunhua 张云华
January, 2015
Written in Beijing

Appendix 1
Chinese Dynasties[4]

Conventional List of Dynasties, Groups of Dynasties, or Periods (Pre-Qin)

Xia 2070–1600 B.C.
Shang c.1600–1046 B.C.
(Yin Shang) c.1300–1046 B.C.
Zhou 1046–256 B.C.
Western Zhou 1046–771 B.C.
Eastern Zhou 770–256 B.C.
Spring and Autumn 722–403 B.C.
Warring States 403–221 B.C.
Six Kingdoms 475-221 B.C.

Conventional List of Dynasties, Groups of Dynasties, or Periods (Imperial China)

Qin 221–206 B.C.
Han
Former (Western) Han 202 B.C.–C.E. 9
Later (Eastern) Han C.E. 27–220
Three Kingdoms 220–280
Jin
Western Jin 265–316
Eastern Jin 317–420

[4]This is a simplified chart based on the list of dynasties and dates found in Endymion Wilkinson, *Chinese History: a New Manual*, pages 3–4. —trans.

Y. Zhang, *Insights into Chinese Agriculture*,
https://doi.org/10.1007/978-981-13-1050-8

Sixteen Kingdoms 304–439
Nanbeichao (Southern and Northern Dynasties) 420–589
Southern Dynasties 420–579
Northern Dynasties 386–581
Sui 581–618
Tang 618–907
Five Dynasties and Ten Kingdoms 907–960
Five Dynasties 920–979
Ten Kingdoms 907–960
Liao Dynasty 916–1125
Song 960–1279
Northern Song 960–1127
Southern Song 1127–1279
Xia (Tangut) 1038–1227
Jin Dynasty (Jurchen) 1115–1234
Yuan 1271–1368
Ming 1368–1644
Qing 1644–1912
The Twentieth Century:
Republic of China 1912–1949
People's Republic of China 1949–

Appendix 2[5]
Weights and Measures

Dry Weights

jin (catty): 500 g or 0.5 kg, equal to 1.1023 pounds
tael: equal to about 50 g, 1/16 of a catty, or 1.763 ounces
ton (*dun*): metric ton or 1,000 kg
shi (picul): equal to about 50 kg
duo (peck): equal to 10 *sheng* to 10 l or 17.60 pints or 2.2 gallons
hu: dry measure, originally equal to 10 *dou*, later (from the Tang dynasty) equal to 5 *dou*

Area

mu [pronounced and romanized as *mou*]: approximately 6 *mu* per acre; 15 *mu* per hectare; or 667 m^2. [One hectare equals 15 *mu* or about 2.471 acres]
fen: a tenth of a *mu* or about 67 m^2
qing: 100 *mu*, equal to 6.667 ha or 16.474 acres

Length

The metric system is throughout this book for linear measurement, with the exception of the mention of the Chinese *li* (or *huali*), known as the Chinese mile. The *li* is traditionally figured as 360 paces and now as one-third of a mile. A *gongli* (a "public *li*") is equivalent to 1 km.

---

[5]This is a simplified chart primarily based on information found in Endymion Wilkinson, *Chinese History: a New Manual*. Wilkinson provides an excellent, detailed discussion of both modern, imperial China, and pre-Qin metrology on pages 551–564. Chinese measurements have been based on the metric system since 1990, though informally the market system is still widely used. See Endymion Wilkinson, *Chinese History: a New Manual*, p. 563. —trans.

Y. Zhang, *Insights into Chinese Agriculture*,
https://doi.org/10.1007/978-981-13-1050-8

Bibliography

*Denotes works not listed in the bibliography of the Chinese version of *Insights into Chinese Agriculture* but mentioned in its text.

**Denotes works added by the translator.

All other works are listed in the bibliography of the Chinese version of this book.

*Akerlof, George. "The Market for Lemons." *Quarterly Journal of Economics*, Vol. 84, No. 3. August 1970.

Alexander Chayanov, A ·恰亚诺夫著, translated by Xiao Zhenghong 萧正洪. *Nongmin jingji zuzhi* 农民经济组织 (*Organization of peasant economy*). Central Compilation Translation Press. 1996.

*Beijing Agricultural University, ed. *Jingji dacidian, nongye jingji juan* 经济大词典, 农业经济卷 (*Dictionary of economics, vol. on agricultural economics*). Shanghai Agricultural Publishing House and Lexicographical Publishing House. 1983.

Brown, Lester R. "Who Will Feed China?" *Worldwatch Magazine*. September, 1994.

Cao Liqun 曹利群, et al. *Chuangxin nongye jingying tizhi jizhi yanjiu* 创新农业经营体制机制研究 (*Research into innovative agricultural economic management systems*). Ministry of Agriculture, *Nongcun jingji wengao* 农村经济文稿 (*Rural economic survey report*), No. 3. 2013.

Chen Dao 陈道, et al. *Jingji dacidian nongye jingji juan* 经济大辞典·农业经济卷 (*Dictionary of economics, section on agricultural economics*). Shanghai Lexicographical Publishing House and China Agriculture Press. 1983.

Chen Xiwen 陈锡文, Zhao Yang 赵阳, Luo Dan 罗丹. *Zhongguo nongcun gaige 30 nian huigu yu zhanwang* 中国农村改革30年回顾与展望 (China's rural reformation: thirty year retrospective and a forecast). People's Publishing House. 2008.

Chen Yiyuan 陈义媛. *Zibenzhuyi shi jiating nongchang de xingqi yu nongye jingying zhuti fenhua de zai sikao* 资本主义式家庭农场的兴起与农业经营主体分化的再思考 (*Rethinking the rise of capitalist-style family farms and the differentiation of farm management*), in *Kaifang shidai* 开放时代 (*Open Times*), No. 4. 2013.

Chen Mengshan 陈萌山. *Zhongguo nongye de feiliao liyong wenti* 中国农业的肥料利用问题 (*Problems of fertilizer utilization in Chinese agriculture*), quoted in Chen Xiaohua, 陈晓华 and Zhang Hongyu 张红宇.*Zhongguo huanjing, ziyuan yu nongye zhengce* 中国环境、资源与农业政策 (China's policies on the natural environment, natural resources, and agriculture). Chinese Agriculture Press. 2006.

Cheng Guoqiang 程国强. *Zhongguo nongye butie zhidu sheji yu zhengce xuanze* 中国农业补贴:制度设计与政策选择 (*China's agricultural subsidies: system designs and government policy choices*) China Development Press. 2011.

***The Chicago Manual of Style, 16th Edition*. University of Chicago Press. 2010.

Engels, Friedrich, ed. *Collected Works of Marx and Engels*. English edition of 1888.

*Er Yuehe 二月河. *Yongzheng wangchao* 雍正王朝 (*The Imperial Court of the Yongzheng Emperor*). Changjiang Literature Publishing Company. Three volumes, 1991, 1993, 1994.

Y. Zhang, *Insights into Chinese Agriculture*,
https://doi.org/10.1007/978-981-13-1050-8

Feng Haifa 冯海发. *Dui shiba jie San Zhong Quanhui "jueding" youguan nongcun gaige jige zhongda wenti de lijie* 对十八届三中全会<决定>有关农村改革几个重大问题的理解 (*Understanding several major issues related to rural reformation in the "Decisions" of the Third Plenary Session of the 18th CPC Central Committee*). Agriculture section of *Nongcun jingji wegao* 农村经济文稿 (*Rural economic survey report*), No. 12. 2013.

General Office of the National Agriculture Commission of the People's Republic of China *Zhonghua renmin gongheguo guojia nongye weiyuanhui bangongting* 中华人民共和国国家农业委员会办公厅, ed. *Nongye jitihua zhongyao wenjian huibian* (*shang, xia ce*) 农业集体化重要文件汇编(上,下册) (*Compilation of important documents pertaining to the collectivization of agriculture, vols.1 and 2*). Party School of the CCP Central Committee Publishing Company. 1981.

Han Jun 韩俊. *Zhongguo shipin anquan baogao* (*2007*) 中国食品安全报告 (2007) (*Report on food safety in China, 2007*). Social Sciences Academic Press (China). 2007.

Han Jun 韩俊, et al. *Zhongguo nongcun gaige* (*2002–2012*) 中国农村改革 (2002—2012). (*China's rural reforms, 2002–2012*). Shanghai Far East Publishers. 2012.

He Anhua 何安华, Zhang Yunhua 张云华. *Woguo liangshi chanhou sunhao jiqi dui liangshi anquan de yingxiang* 我国粮食产后损耗及其对粮食安全的影响 (*Post-production shrinkage of China's staple foods, as well as its influence on food security*). *Zhongguo wujia* 中国物价 (*Chinese commodity prices*), No. 6. 2013.

Hu Junhua 胡军华 and Chen Jiayi 陈佳怡. *Zhongguo liangshi jiage cheng shijie "gaodi"* 中国粮食价格成世界"高地" ("China's staple foods prices have become the world's 'highlands'"). *Di-yi caijing ribao* 第一财经日报 (*China Business News* [Shanghai]). July 29, 2014.

Huang, Philip C. C. 黄宗智. *Zhongguo de yingxing nongye geming* 中国的隐形农业革命 (China's hidden agricultural revolution). Law Press. 2010.

Huang, Philip C. C. 黄宗智. *Zhongguo xiaonong jingji de guoqu he xianzai—shuerci lilun de duicuo*中国小农经济的过去和现在—舒尔茨理论的对错 (*China's small farm economy in the past and present— correct and incorrect aspects of Schultz's theory*). Zhongguo xiangcun yanjiu中国乡村研究 (Research into Chinese rural areas), No. 6. Fujian Education Press. 2008.

Jiang Heping 蒋和平, et al. *Fazhan nongye shidu guimo jingying de shixian lujing yu zhengce jianyi* 发展农业适度规模经营的实现路径与政策建议 (*Policy proposals and methods to develop appropriately large-scaled agricultural management*). *Nongyebu* (Ministry of Agriculture), *Nongcun jingji wegao*农村经济文稿 (*Documents on rural economics*), No. 12. 2012.

Li Hengquan 李恒全. *Han-chu xiantianzhi he tianshui zhengshou fangshi* 汉初限田制和田税征收方式 (*The limited field system and methods of land taxes collection in the early Han dynasty*). *Zhongguo jingjishi yanjiu* 中国经济史研究 (*Research in Chinese economic history*), No. 1. 2007.

Liu Jianshui 刘建水. *Guowai xiandai jiating nongchang de jiben tezheng* 国外现代家庭农场的基本特征 (*Fundamental characteristics of overseas modern family farms*), in *Nongcun jingji wengao* 农村经济文稿 (*Documents on the Rural Economy*) of *Nongyebu* (*Ministry of Agriculture*), No. 12. 2013.

Liu Qi 刘奇. *Nongdi kunju* 农地困局 (*Dilemmas facing farmland*). *Zhongguo fazhan guancha* 中国发展观察 (*China development observation*), 2014, No. 6.

Liu Ruibai 刘瑞柏. *Zhongguo liangshi wuliu hangye fazhan qianjing yu touzi yuce fenxi baogao qianzhan* 中国粮食物流行业发展前景与投资预测分析报告前瞻 ("*Analysis of prospects of China's food staples logistics industry and investment forecast*"). *Qianzhan zixun* 前瞻资讯 (*Forecast information*). 2013.

Liu Ti 刘逖 *1600-1840 nian Zhongguo guonei shengchan zongzhi de gusuan* 1600—1840 年中国国内生产总值的估算 (*An estimate of China's gross national product from 1600 to 1840*). *Jingji yanjiu* 经济研究 (*Economic Research Journal*), No. 10. 2009.

Luo Xun 罗荀. *Rushi shi-zhounian tegao* 入世10周年特稿 ("Feature article for commemorating the tenth anniversary of entering the World Trade Organization"). *Zhongguo caijing bao* 中国财经报 (*China Financial and Economic News*). December 8, 2011.

*Ni Tao 倪涛. *Qihou zhinengxing nongye jianfu shuangchong shiming* 气候智能性农业肩负双重使命 ("Dual Missions of Farming: Climate-Smart Agriculture"). *Renmin Ribao* 人民日报 (*People's Daily*). December 6, 2013.

Research Group of the Department of Industrial Policy and Regulation of the Ministry of Agriculture, Nongyebu chanye zhengce yu fagui si diaoyanzu 农业部产业政策与法规司调研组. *Shanghai Songjiang jiating nongchang diaocha baogao* 上海松江家庭农场调查报告 (*Survey report on family farms in the Songjiang district of Shanghai*). *Nongcun jingji wengao* 农村经济文稿 (*Rural economic survey report*) of the *Nongyebu* (Ministry of Agriculture), No. 6. 2013.

**National Bureau of Statistics of the People's Republic of China, Zhonghua renmin gongheguo guojia tongjiju* 中华人民共和国国家统计局. *Zhongguo tongji nian jian [1981–2014]* 中国统计年鉴 (1981–2014) (*China Statistical Yearbook*). 1981 through 2014.

**Qin Hui 秦晖. '*Nongmin jianfu*' *yao fangzhi Huang* Zongxi dinglü xianjing '农民减负'要防止黄宗羲定律陷阱 ("When lightening the burden on farmers, beware of the trap of Huang Zongxi's Law"). Zhongguo jingji shibao 中国经济时报 (Chinese Economic Times). March 21, 2001.

Schultz, Theodore W 西奥多·W·舒尔茨著, translated by Liang Xiaomin 梁小民. *Gaizao chuantong nongye* 改造传统农业 (Transforming traditional agriculture). The Commercial Press (Beijing). 2003. [Chinese translation of: Schultz, Theodore. *Transforming Traditional Agriculture.*Yale University Press. 1964.]

*Schumacher, E. F. *Small Is Beautiful*: *A Study of Economics As If People Mattered.* Blond & Briggs (1973–2010), HarperCollins (2010–present).

Shōgenji Shin'ichi 生源寺真一. *Riben de shiwu he nongye zhengce*日本的食物和农业政策 (*Japan's food and agricultural policies*). *Zhongguo nongye daxue jingguan xueyuan* 中国农业大学经管学院 (School of Economics and Management, China Agricultural University), "*Riben zhongce yantaohui*" *huiyi cailiao* "日本政策研讨会"会议材料 (Conference materials for the "Seminar on Japanese policies"). May, 2014.

Research Project Group, State Council Research Office *Guowuyuan yanjiushi ketizu* 国务院研究室课题组. *Zhongguo nongmin gong diaoyan baogao* 中国农民工调研报告 (*Survey report on migrant workers in China*). China Yan Shi Press. 2006.

Sun Honglei 孙洪磊 and Zhang Zhilong 张志龙. *Jiating nongchang "qiangzhu chao" ancang yinyou* 家庭农场"抢注潮"暗藏隐忧 ("What lies behind the frantic rush to register as a 'family farm'"). *Jingji cankao bao* 经济参考报 (*Economic Information Daily*), No. 5. January 3, 2014.

*Sun Zhilu 孙之騄 (fl. Qing dynasty). *Er shen yelu* 二申野录 (*Unofficial history of 1368–1644* [*i.e., the Ming Dynasty*]). Shandong Qilu Press. 1996.

Van der Ploeg, Jan Douwe 杨·杜威·范德普勒格著, translated by Pan Lu 潘璐, Ye Jingzhong 叶敬忠, et al. *Xin xiaonong jieji* 新小农阶级 (*The New Peasantries*). Social Sciences Academic Press [China]. 2013. [Chinese translation of: van der Ploeg, Jan Douwe. *The New Peasantries*: *Struggles for Autonomy and Sustainability in an Era of Empire and Globalization.* London and Sterling, Virginia. Earthscan. 2008.]

Xiao Junyan 肖俊彦. *Zhongguo nongchanpin gongqiu de hexin—liangshi anquan wenti* 中国农产品供求的核心——粮食安全问题 (*At the core of the supply and demand of Chinese agricultural products—the food* [liangshi] *security issue*). *Zhongguo tese nongye xiandaihua daolu yanjiu* 中国特色农业现代化道路研究 (*Research into the path to agricultural modernization with Chinese characteristics*), in *Guowuyuan fazhan yanjiu zhongxin nongcun jingji yanjiubu ketizu* 国务院发展研究中心农村经济研究部课题组 (*Research group of the rural economy research department, the Development Research Center of the State Council*). China Development Press. 2012.

Xu Wei 徐伟 and Zhao Yangzi 赵杨子. *Rang baixing "shide quan" gengyao "shide an"—fang guowuyuan fazhan yanjiu zhongxin nongcun jingji yanjiubu fu yanjiuyuan Zhang Yunhua* 让百姓"食得全"更要"食得安"——访国务院发展研究中心农村经济研究部副研究员张云华 ("Provide people with varied and safe foods—interview with associate research fellow Zhang

Yunhua of the Rural Economy Research Department in the Development Research Center of the State Council"). *Zhongguo jingji shibao* 中国经济时报 (*China Economic Times*). September 21, 2011.

*Wang Yuewen 王跃文. *Da Qing xiangguo* 大清相国 (*Prime minister of the Great Qing dynasty*). Hunan Literature and Art Publishing House. 2013.

*Wang Zhongluo 王仲荦. *Sui,Tang, Wudai shi (shang)* 隋唐五代史（上）(*History of the Sui, Tang, and Five Dynasties, vol.1*). Shanghai People's Publishing House. 2003.

Wang Zuli 王祖力. *Zhongguo huafei shichang yu zhengce yanjiu* 中国化肥市场与政策研 (*Research into the Chinese fertilizer market and government policy*). China Agriculture Press. 2011.

**Wilkinson, Endymion. *Chinese History: A New Manual, 5th edition*. Harvard University Asian Center. 2018.

Ye Jingzhong 叶敬忠. *Meiyou xiaonong de shijie hui hao ma?—jianxu "Xin xiaonong jieji" zhongyiben* 没有小农的世界会好吗?——兼序《新小农阶级》中译本 (*Would the world be better off without small farmers?—Chinese translation of the introduction* The New Peasantries [*by Jan Douwe van der Ploeg*]). *Zhongguo renmin daxue shubao ziliao zhongxin* 中国人民大学书报资料中心 (*Renmin University of China bibliographic data center*), *Nongye jingji yanjiu* 农业经济研究 (*Agricultural economy*), No. 11. 2013.

*Ye Shengtao 叶圣陶. *Duo shou le san-wu dou* 多收了三五斗 (*This harvest is greater by three to five pecks*). Short story first published in *Wenxue* 文学 (*Literature*). July 1, 1933.

Yoshio Kobayashi 小林芳雄. *Guanyu Riben nongye de zuijin dongxiang* 关于日本农业的最近动向 (Recent trends in Japanese agriculture). *Zhongguo nongye daxue jingguanxueyuan* 中国农业大学经管学院 (China Agricultural University, School of Economics and Management), "*Riben zhengce yantaohui*" *huiyi cailiao* "日本政策研讨会"会议材料 (Materials used in the "Japanese Government Policy Seminar"). May, 2014.

Yuan Guoming 袁国铭. *Shipin shengchan anquan kongzhi he zhengfu jianguan dou hen zhongyao—zhuanfang guowuyuan fazhan yanjiu zhongxin nongcun jingji yanjiubu fu yanjiuyuan Zhang Yunhua* 食品生产安全控制和政府监管都很重要——专访国务院发展研究中心农村经济研究部副研究员张云华 ("Safety control and government supervision are both equally important in food production—a special interview with associate research fellow Zhang Yunhua of the Rural Economic Research Department in the Development Research Center of the State Council"). *Zhongguo jingji shibao* 中国经济时报 (*China Economic Times*). July 27, 2012.

Zhang Jin'guang 张金光. *Pubian shoutianzhi de zhongjie yu siyoudiquan de xingcheng* 普遍授田制的终结与私有地权的形成 ("*The end of the universal land allotment systems and the formation of private land ownership rights*"). *Lishi yanjiu* 历史研究 (*Historical research*), No.5. 2007.

Zhang Yunhua 张云华. *Shipin anquan baozhang jizhi yanjiu* 食品安全保障机制研究 (*Research into safeguard mechanisms of food safety*). China Water and Power Press. 2007.

Zhang Yunhua 张云华 and Yang Xiaoyan 杨晓艳. *Xin nongcun jianshe zhong de nongcun zhiye jiaoyu yu nongmin peixun* 新农村建设中的农村职业教育与农民培训 (*Rural vocational education and famer training in construction of the new countryside*). *Beijing nongye zhiye xueyuan xuebao* 北京农业职业学院学报 (*Journal of Beijing Vocational College of Agriculture*), No. 3. 2007.

Zhang Yunhua. *Jin yibu hangshi liangshi anquan de zhichi tixi* 进一步夯实粮食安全的支持体系 (*Further strengthen the foundations of the food security support system*). *Guowuyuan fazhan yanjiu zhongxin diaocha yanjiu baogao zeyao* 国务院发展研究中心调查研究报告择要 (*Abstracts of investigative reports of the Development Research Center of the State Council*), No. 34. 2008.

Zhang Yunhua. *Henan sheng liangshi wenti diaoyan baogao* 河南省粮食问题调研报告 (*Investigative report on the issue of staple foods in Henan Province*). *Guowuyuan fazhan yanjiu zhongxin diaocha yanjiu baogao* 国务院发展研究中心调查研究报告 (*Investigative reports of the Development Research Center of the State Council*), No. 81. 2008.

Zhang Yunhua. *Guanyu zhiding 'nongcun jiti jingji zuzhi fa' de sikao* 关于制定"农村集体经济组织法"的思考 ("Thoughts on formulating the Rural Collective Economic Organizations Law"). *Nongye jingji wenti* 农业经济问题 (*Problems of agricultural economy*), No. 5. 2010.

Zhang Yunhua *Nongmin tudi jingying quan diya daikuan zhide tuiguang—Ningxia Tongxin xian tudi diya daikuan diaocha baogao* 农民土地经营权抵押贷款值得推广—宁夏同心县土地抵押贷款调查报告 (*Mortgaging rural land management rights is worth promoting—investigative reports on the mortgaging of land rights in Tongxin County, Ningxia Hui Autonomous Region*). *Guowuyuan fazhan yanjiu zhongxin diaocha yanjiu baogao* 国务院发展研究中心调查研究报告 (*Investigative reports of the Development Research Center of the State Council*), No. 163. 2010.

Zhang Yunhua, et al. *Wanshan yu gaige nongcun zhaijidi zhidu yanjiu* 完善与改革农村宅基地制度研究 (*Research into the refinement and reform of the rural homesite system*). China Agriculture Press. 2011.

Zhang Yunhua, et al. *Tongchou chengxiang fazhan zhong nongmin tudi quanyi baozhang yanjiu* 统筹城乡发展中农民土地权益保障研究 (*Research into safeguarding land ownership rights and interests of farmers in balancing urban and rural development*). *Nongyebu ruankexue keti baogao* 农业部软科学课题报告 (*Ministry of Agriculture project report on soft science*). 2011.

Zhang Yunhua, et al. *Zhongguo nongdi liuzhuan wenti diaocha* 中国农地流转问题调查 (*Investigation into farmland transaction issues in China*). Shanghai Far East Publishers. 2012.

Zhang Yunhua, et al. *Nongmin tudi caichan quanli wenti yanjiu* 农民土地财产权利问题研究 (*Research into issues of farmers' land and property rights*). *Nongyebu ruankexue keti baogao* 农业部软科学课题报告 (*Ministry of Agriculture project report on soft science*). 2012.

Zhang Yunhua. *Jinyibu hangshi nongmin de tudi caichan quanli* 进一步夯实农民的土地财产权利 ("Further strengthen land and property rights of farmers"). *Nongmin ribao* 农民日报 (*Farmers' Daily*), page 3. October 1, 2013.

Zhang Yunhua. *Ruhe yingdui "shigao" nongye de tiaozhan—Zhongguo nongye fazhan de jieduanxing tezheng he zhengce quxiang* 如何应对"十高"农业的挑战——中国农业发展的阶段性特征和政策取向 ("How to respond to the challenges of 'Ten High' agriculture—phasic characteristics and policy orientation of Chinese agricultural development"). *Nongmin ribao* 农民日报 (*Farmers' Daily*), page 3. August 20, 2013.

Zhang Yunhua and Guo Cheng 郭铖. *Nongye jingying tizhi chuangxin de Jiangsu ge'an: tudi gufen hezuo yu shengchan zhuanye chengbao* 农业经营体制创新的江苏个案:土地股份合作与生产专业承包 (*The case of Jiangsu Province in creative agricultural management systems: joint land shares and specialized production responsibility contracts*). *Gaige* 改革(*China Reform*), No.2. 2013.

Zhao Gang 赵冈 and Chen Zhongyi 陈钟毅. *Zhongguo tudi zhidu shi* 中国土地制度史 (*History of Chinese land tenure systems*). New Star Press. 2006. [Originally published by Liangting Publishing House, Taipei. 1982.]

*Zhao Rong 赵荣. *Renwen Shaanxi* 人文陕西 (*Man and Culture in Shaanxi Province*). Shaanxi Travel and Tourism Press. 2010.

